MAKING SENSE OF MATHEMATICS FOR TEACHING

High School

EDWARD C. NOLAN
JULI K. DIXON
FARSHID SAFI
ERHAN SELCUK HACIOMEROGLU

Solution Tree | Press
a division of Solution Tree

555 North Morton Street
Bloomington, IN 47404
800.733.6786 (toll free) / 812.336.7700
FAX: 812.336.7790

email: info@solution-tree.com
solution-tree.com

Visit **go.solution-tree.com/mathematics** to access materials related to this book.

Printed in the United States of America

20 19 18 17 16 1 2 3 4 5

FSC
www.fsc.org
MIX
Paper from responsible sources
FSC® C011935

Library of Congress Control Number: 2016933463

ISBN: 978-1-942496-48-9

Solution Tree
Jeffrey C. Jones, CEO
Edmund M. Ackerman, President

Solution Tree Press
President: Douglas M. Rife
Senior Acquisitions Editor: Amy Rubenstein
Managing Production Editor: Caroline Weiss
Senior Production Editor: Rachel Rosolina
Proofreader: Ashante K. Thomas
Text and Cover Designer: Abigail Bowen

Acknowledgments

Many thanks to my family, Michele and Calvin, for the wonderful support they continue to provide me in all of my endeavors. Many thanks to the Dixon Nolan Adams Mathematics associates who worked as a collaborative team in providing feedback on the content of this book. And finally, thanks to Jeff Jones, Douglas Rife, and Stefan Kohler from Solution Tree for helping us and believing in us and our vision for making sense of mathematics for teaching.

—Ed Nolan

My deepest love and gratitude to my daughters, Alex and Jessica, who continue to allow me to understand mathematics more deeply through their eyes, and to my husband, Marc, who supports me to spend countless hours doing so. I also wish to thank our author team—Ed, Farshid, and Selcuk—for the wonderful collaborative environment they supported while writing this book.

—Juli Dixon

I would like to thank my wife, Farinaz, and our two wonderful kids, Neeka and Neema, for their patience, love, and support. To my mother, Pouran, and my father, S. Ali Safi, the two of you have been my role models, source of strength, and compass in life's miraculous journey. To my siblings, Omid, Farnaz, and Farzad, you continue to amaze me with your talents and pursuit of excellence. I would also like to express my sincere gratitude to my dear colleagues Juli, Ed, and Selcuk for their vision, guidance, and commitment to make this dream a reality.

—Farshid Safi

I would like to thank my wife, Isomi, for her love, patience, and support. To my children, Ayla, Deniz, and Selin, who inspire me and keep me focused on the most important things in my life. To my mother, Nefise, who supported and believed in me even when I didn't. To my sisters Senem and Guney, thank you for always being so kind and supportive. To my dear colleagues Juli, Ed, and Farshid, I cannot thank you enough for all your support, guidance, and encouragement.

—Selcuk Haciomeroglu

Solution Tree Press would like to thank the following reviewers:

Shelby Aaberg
Mathematics Teacher
Scottsbluff High School
Scottsbluff, Nebraska

Sue Abrams
Mathematics Teacher and PLC Leader
Montpelier High School
Montpelier, Vermont

Michael Conklin
Mathematics Teacher
University High School
Spokane Valley, Washington

Brian Dean
Senior Instructional Specialist, 6–12 Mathematics
District School Board of Pasco County
Land O' Lakes, Florida

Noah Franske
Mathematics Teacher
Edina High School
Edina, Minnesota

Elisabeth Jaffe
Mathematics Teacher
Baruch College Campus High School
New York City, New York

Chance Nalley
Mathematics Educator
Horace Mann School
Bronx, New York

Jennifer Wilson
Mathematics Teacher
Northwest Rankin High School
Flowood, Mississippi

Visit **go.solution-tree.com/mathematics** to access live links to the websites mentioned in this book.

Table of Contents

About the Authors

Edward C. Nolan is the preK–12 director of mathematics for Montgomery County Public Schools in Maryland. He has nineteen years of classroom experience in both middle and high schools and was department chair for fifteen years, all in Montgomery County. An active member of the National Council of Teachers of Mathematics (NCTM), he is currently the president of the Maryland Council of Supervisors of Mathematics. Nolan is also a consultant for Solution Tree as one of the leaders of Dixon Nolan Adams Mathematics, providing support for teachers and administrators on the rigorous standards for mathematics.

Nolan has been published in the *Banneker Banner*, a publication of the Maryland Council of Teachers of Mathematics, and *Mathematics Teaching in the Middle School*, an NCTM publication, and he has conducted professional development at the state, regional, and national level, including webinars for NCTM and TODOS: Mathematics for ALL. His research interests lie in helping students and teachers develop algebraic thinking and reasoning. In 2005, Nolan won the Presidential Award for Excellence in Mathematics and Science Teaching.

He is a graduate of the University of Maryland. He earned a master's degree in educational administration from Western Maryland College.

To learn more about Nolan's work, follow @ed_nolan on Twitter.

Juli K. Dixon, PhD, is a professor of mathematics education at the University of Central Florida (UCF) in Orlando. She coordinates the award-winning Lockheed Martin/UCF Academy for Mathematics and Science for the K–8 master of education program as well as the mathematics track of the doctoral program in education. Prior to joining the faculty at UCF, Dr. Dixon was a secondary mathematics educator at the University of Nevada, Las Vegas and a public school mathematics teacher in urban school settings at the elementary, middle, and secondary levels.

She is a prolific writer who has authored and coauthored books, textbooks, chapters, and articles. A sought-after speaker, Dr. Dixon has delivered keynotes and other presentations throughout the United States. She has served as chair of the National Council of Teachers of Mathematics Student Explorations in Mathematics Editorial Panel and as a board member for the Association of Mathematics Teacher Educators. At the state level, she has served on the board of directors for the Nevada Mathematics Council and is a past president of the Florida Association of Mathematics Teacher Educators.

Dr. Dixon received a bachelor's degree in mathematics and education from the State University of New York at Potsdam, a master's degree in mathematics education from Syracuse University, and a doctorate in curriculum and instruction with an emphasis in mathematics education from the University of Florida. Dr. Dixon is a leader in Dixon Nolan Adams Mathematics.

To learn more about Dr. Dixon's work supporting children with special needs, visit www.astrokeofluck.net or follow @thestrokeofluck on Twitter.

Farshid Safi, PhD, is an assistant professor of mathematics education at the University of Central Florida (UCF) in Orlando. He focuses on developing teachers' conceptual understanding of elementary and secondary mathematics while connecting essential mathematical topics through the use of multiple representations and technology. Prior to joining the faculty at UCF, Dr. Safi was a faculty member in the Department of Mathematics and Statistics at The College of New Jersey for six years. As a public school mathematics teacher in Florida, he taught courses ranging from algebra to Advanced Placement calculus. In addition, he has taught mathematics and mathematics education content and methods courses at several institutions such as the University of Florida, University of North Florida, University of British Columbia, and Simon Fraser University in Canada.

Dr. Safi has been published in *Mathematics Teaching in the Middle School*, an NCTM publication, as well as other national and international journals. He has presented at state, national, and international conferences and has conducted professional development sessions throughout the United States focusing on the teaching and learning of mathematics. He is a member of the National Council of Teachers of Mathematics and the Association of Mathematics Teacher Educators (state and national levels) and has served in leadership positions for both organizations.

Dr. Safi earned a bachelor's degree and a master's degree in mathematics with a focus on teaching from the University of Florida and a doctorate in education with an emphasis on mathematics from the University of Central Florida.

To learn more about Dr. Safi's work, follow @FarshidSafi on Twitter.

Erhan Selcuk Haciomeroglu, PhD, is an associate professor of mathematics education at the University of Central Florida, where he also coordinates the undergraduate and graduate mathematics education programs. His research focuses on the conceptual understanding of calculus and technology integration in teaching and teacher education.

During his graduate studies at Florida State University, Dr. Haciomeroglu taught several undergraduate courses for the Department of Mathematics and the Mathematics Education Program in the Department of Middle and Secondary Education. He was also a high school mathematics teacher in Turkey for eight years, teaching a variety of mathematics courses ranging from arithmetic to calculus.

He has written articles and delivered presentations at international, national, and state conferences on teaching and learning calculus and technology integration in the mathematics classroom. He has also served as a member on the Florida Council of Teachers of Mathematics board of directors.

Dr. Haciomeroglu received a bachelor's degree in mathematics from Çukurova University and master's and doctoral degrees in mathematics education from Florida State University.

To learn more about Dr. Haciomeroglu's work, follow @eselcukh on Twitter.

To book Edward C. Nolan, Juli K. Dixon, Farshid Safi, or Erhan Selcuk Haciomeroglu for professional development, contact pd@solution-tree.com.

Introduction

The only way to learn mathematics is to do mathematics.

—Paul Halmos

When teaching, much of the day is spent supporting students to engage in learning new content. In mathematics, that often means planning for instruction, delivering the planned lessons, and engaging in the formative assessment process. There are opportunities to attend conferences and other professional development events, but those are typically focused on teaching strategies or on administrative tasks like learning the new gradebook program. Opportunities to take on the role of *learner* of the subject you teach are often neglected. As you read *Making Sense of Mathematics for Teaching High School*, you will have the chance to become the learner once again. You will *learn* about the mathematics you teach by *doing* the mathematics you teach.

There is a strong call to build teachers' content knowledge for teaching mathematics. A lack of a "deep understanding of the content that [teachers] are expected to teach may inhibit their ability to teach meaningful, effective, and connected lesson sequences, regardless of the materials that they have available" (National Council of Teachers of Mathematics [NCTM], 2014, p. 71). This lack of deep understanding may have more to do with lack of exposure than anything else.

All too often, exposure to mathematics is limited to rules that have little meaning. Teachers then pass these rules on to students. For example, in the early years of high school, the way mathematics is taught influences students' understanding of advanced mathematics. A teacher might teach students that they can determine whether a relationship is a function by using the vertical line test. While this is an effective method for making such a determination regarding a mathematical relation, teachers need to be sure to teach the properties of function with depth, or students will have only a superficial understanding of this necessary topic. When students move to studying inverse functions that may involve restricted domains, they need to understand how functions require one output element—and only one output element—for every input element. This may not be well understood if the only way that students understand the definition of a function is through the vertical line test. It might be that the vertical line test was how you were taught, but it is not sufficient to pass on to students as the best (or only) way to define a function.

This book is our response to requests from teachers, coaches, supervisors, and administrators who understand the need to know mathematics for teaching but who don't know how to reach a deeper level of content knowledge or to support others to do so. First and foremost, the book provides guidance for refining what it means to be a teacher of mathematics. To teach mathematics for depth means to facilitate instruction that empowers students to develop a deep understanding of mathematics. This can happen when teachers are equipped with strong mathematics content knowledge—knowledge that covers the conceptual understanding and procedural skill of mathematics and knowledge that is supported by a variety of strategies and tools for engaging students to learn mathematics. With these elements as a backdrop, this book can be used to go below the surface in core areas of mathematics.

Second, coaches, supervisors, and administrators benefit from the content and perspectives provided in this book because it offers a source that supports guidance and mentoring to enhance teachers' mathematics content knowledge and their knowledge for teaching mathematics. They can particularly benefit from this book as a resource for helping them recognize expected norms in mathematics classrooms.

Here, we will set the stage for what you will learn from this book along with the rationale for why it is important for you to learn it. First, we provide some of the reasons why teachers need to understand mathematics with depth. Next, we share the structure of each chapter along with a description of what you will experience through that structure. Finally, we present ways that you will be able to use this book as an individual or within a collaborative team.

A Call for Making Sense of Mathematics for Teaching

Often, teachers are not initially aware that they lack sufficient depth of mathematical understanding or that this depth of understanding is critical to being equipped to guide students' mathematical development. What we have found is that engaging in tasks designed to contrast procedural and conceptual solution processes provides a window into the gap left by teaching and learning mathematics without understanding.

Procedural skill includes the ability to follow rules for operations with a focus on achieving a solution quickly, while *conceptual understanding* includes comprehension of mathematical ideas, operations, and relationships. The procedure for identifying the slope-intercept form of a linear function is one that teachers can execute without much thought; whereas, the conceptual understanding needed to compose a word problem to represent the same problem might be less accessible. Thus, the contrast between typical solution processes and those that develop conceptual understanding highlights the need to truly comprehend mathematics in a deep way in order to teach it.

As a team, we provide large-scale professional development workshops for school districts across the United States and beyond. We often begin our presentations by engaging participants in a short mathematical activity to set the stage for the types of mind-shifting approaches necessary to teach for depth. One such activity involves linking the steps in equation solving to a graphical representation of those steps (see figure I.1).

Participants frequently are familiar with the algebraic process on the left but are not as familiar with the connection to the graphical representation on the right. How does the equation relate to the graph? How does the graph help develop understanding of the equation? The participants begin to make the connections between the two approaches, and while they understand the mathematics involved, many respond that they had not previously considered how the equation-solving process connected to the graphical representation and how the solution of the equation would appear in a graph of each step. That is a comment we often hear in regard to some of the shifts around developing a deep understanding of mathematical content—that these strategies make sense but feel entirely new in many ways. Making sense of problems in this way is likely not the way you learned mathematics.

Consider the two different approaches to solving the equation $x^2 - 5x + 4 = 2(x - 1)$.

How are the approaches similar? How are they different?

Approach 1

$x^2 - 5x + 4 = 2(x - 1)$

$x^2 - 5x + 4 = 2x - 2$

$x^2 - 7x + 4 = -2$

$x^2 - 7x + 6 = 0$

$(x - 6)(x - 1) = 0$

$x = 6$ and $x = 1$

Approach 2

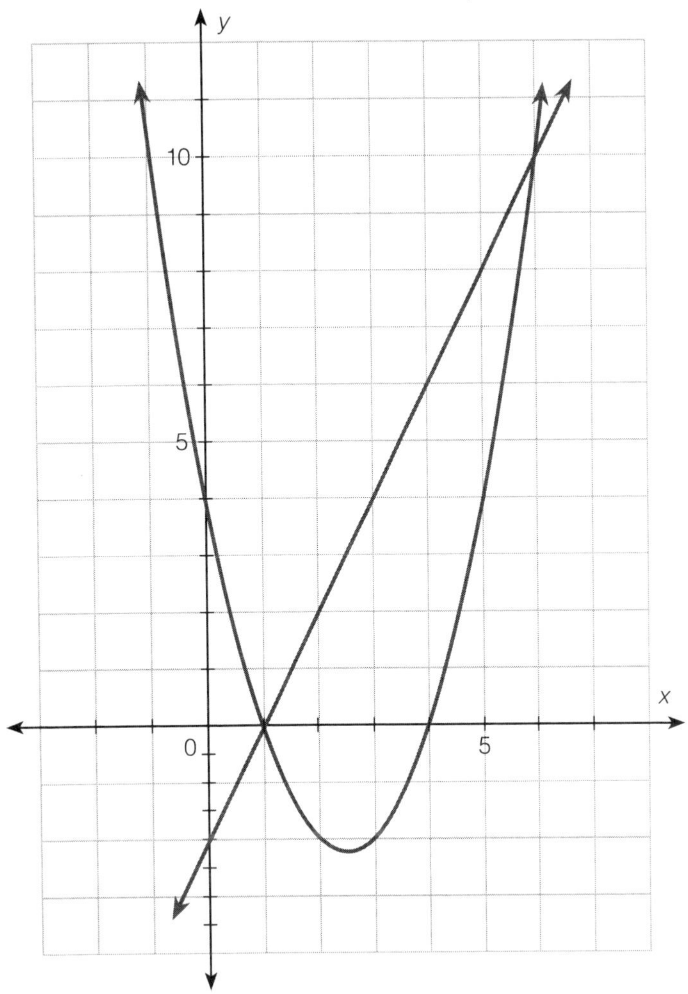

Figure I.1: Solving an equation using an algebraic and a graphical approach.

One of the keys to making the connection between the algebraic and graphical approaches is understanding what each process represents. The algebraic process represents using properties of equality to generate equivalent equations that move toward determining a value for the variable. Isolating the variable in this way allows you to determine the value or values that make the equation true. The graphical approach involves representing the expressions that make up the equation as separate functions that can be graphed. The *x*-values of the intersection of these two graphs represent the set of solutions to the equation. The connection between these approaches helps you see that the solution or solutions to a pair of algebraic functions can be represented as a point, or points, on a graph, signifying how a collection of points on a single graph represents the set of solutions to a single equation.

How would graphs of later steps in the equation-solving process appear, such as for $x^2 - 7x + 4 = -2$? Another interesting facet of the connection between these approaches is how the graphical representation changes through the steps of the algebraic solution process. What do you predict will be the same and what will be different when the next stage in the algebraic solution is represented graphically? See figure I.2.

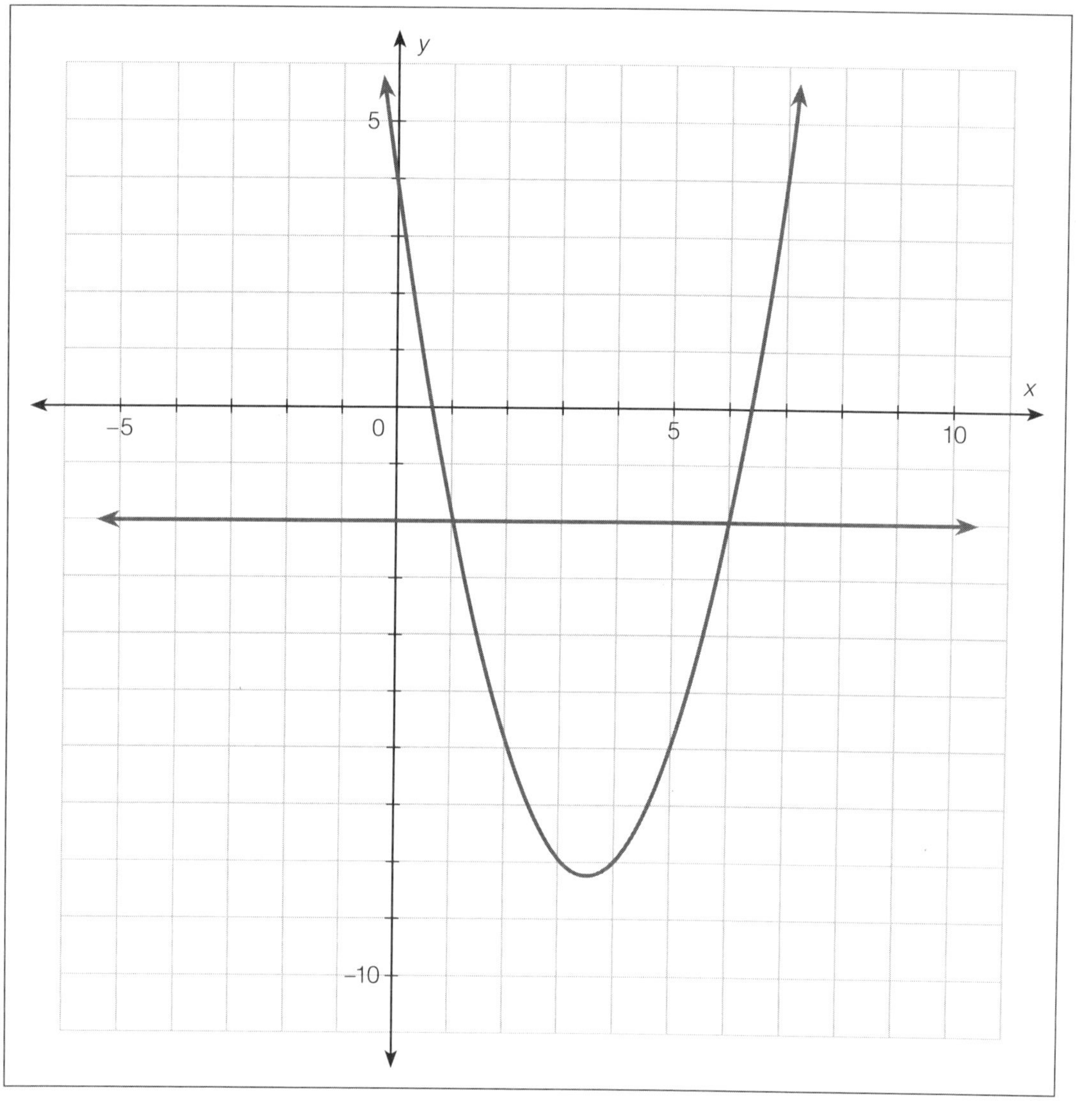

Figure I.2: Graphical representation of $x^2 - 7x + 4 = -2$.

Note that the x-values of the intersections between the two graphs are still 1 and 6, but the y-values have changed. The values of the points of intersection have changed to correspond to the changes from applying the properties of equality in order to work toward isolating the variable in the algebraic procedure. What do you predict will happen if $x^2 - 7x + 6 = 0$ is graphed? Figure I.3 provides the graph of the solution at this stage in the algebraic manipulation.

Notice how the graph now provides two different ways to consider the solution to the original equation. The points of intersection for $y = x^2 - 7x + 6$ and $y = 0$ are (1, 0) and (6, 0); the x-coordinates indicate the

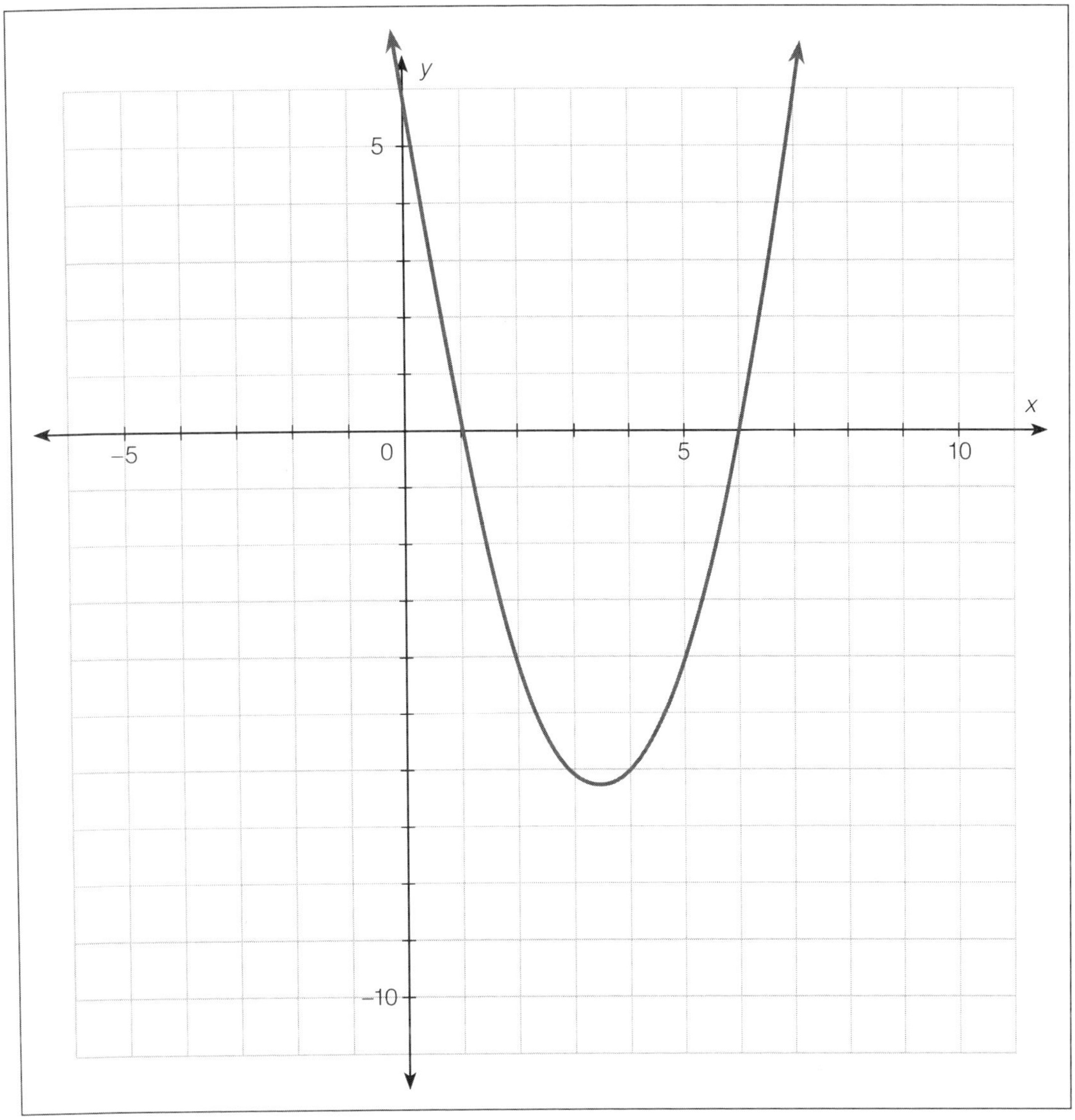

Figure I.3: Graphical representation of $x^2 - 7x + 6 = 0$.

solution to the equation. In addition, the *x*-intercepts of graphing the left side of the equation also provide the solution. This is a representation that many algebra teachers understand. It is, in fact, a representation that many learned as a way to graphically represent the solution. It is the original and intermediate steps of processes similar to the one just described that rarely are represented when considering a graphical solution to solving an equation.

Understanding the connections between different representations as they change throughout the solution process represents one of the key shifts a deep understanding of mathematics will foster. Teachers who do not have this depth of understanding may miss opportunities to develop it in their students. They must also understand how content is developed from one year to the next, use effective mathematics practices to build mathematical proficiency in all students, and incorporate tasks, questioning, and evidence into instruction.

Understanding Mathematics for Teaching

How does one develop the ability to flexibly move between algebraic and graphical approaches? Even if *you* have this understanding, how do you help your fellow teachers or the teachers you support develop this depth of knowledge? Questions such as these led us to create our mathematics-content-focused professional development institutes and the accompanying follow-up workshops, in which teachers implement new skills and strategies learned at the mathematics institutes. Conversations during the follow-up workshops provide evidence that teachers benefit from knowing the mathematics with depth, as do the students they serve. After all, discussing the connection between algebraic and graphical approaches within teacher teams is a powerful way to develop a deep understanding.

We begin each follow-up workshop with a discussion of what is going well at the participants' schools and what needs further attention. Their responses to both queries reaffirm our need to focus on teachers' pedagogical content knowledge. A typical response regarding what is going well includes a discussion about how teachers are now able to make mathematical connections between the topics they teach. Teachers begin to recognize commonalities between topics, such as the connection between quadratic functions and conic sections. They might explore the relationship between completing the square when solving a quadratic equation and determining the maximum value of that equation in context. Teachers report that, in past years, they taught quadratic functions and conic sections as separate topics, making few connections between them, as though the topics existed in silos, completely separate. They taught without coherence. With a deeper understanding of the content they teach, however, they note that they are able to reinforce earlier topics and provide rich experiences as they make connections from one topic to the next. Similarly, coaches report that their deeper mathematics understanding is useful in helping teachers attend to these connections during planning and instruction and within the formative assessment process. The formative assessment process includes the challenging work of evaluating student understanding throughout the mathematics lesson and unit. Teachers need a deep understanding of the mathematics they teach to support a thoughtful process of making sense of student thinking and being confident to respond to students' needs whether those needs include filling gaps, addressing common errors, or advancing ideas beyond the scope of the lesson or unit.

Through the mathematics institutes and workshops, participants realize the need for additional professional development experiences, but providing this level of support can be challenging for schools and districts. This book is our response to this need. We've designed it to support stakeholders who want a review as well as to address additional topics. Our approach herein is informed by our extensive experience providing professional development throughout the United States as well as internationally and is supported by research on best practices for teaching and learning mathematics.

Engaging in the Mathematical Practices

As teachers of mathematics, our goal for all students should be mathematical proficiency, regardless of the standards used. One way to achieve mathematical proficiency is to "balance *how* to use mathematics with *why* the mathematics works" (National Council of Supervisors of Mathematics [NCSM], 2014, pp. 20–21, emphasis added). Mathematical proficiency involves unpacking the mathematics embedded within learning progressions, developing and implementing an assortment of strategies connected to mathematical topics and the real world, being able to explain and justify mathematical procedures, and interpreting and making sense of students' thinking (Ball, Thames, & Phelps, 2008). These processes are well described by the eight Standards for Mathematical Practice contained within the Common Core State Standards (CCSS) for mathematics (National Governors Association Center for Best Practices [NGA] & Council of Chief State School Officers [CCSSO], 2010).

1. Make sense of problems and persevere in solving them.
2. Reason abstractly and quantitatively.
3. Construct viable arguments and critique the reasoning of others.
4. Model with mathematics.
5. Use appropriate tools strategically.
6. Attend to precision.
7. Look for and make use of structure.
8. Look for and express regularity in repeated reasoning.

The Mathematical Practices describe the ways that mathematically proficient students solve problems and engage in learning mathematics. What does this mean for you and your students? Since the Mathematical Practices truly describe how *students* engage with the mathematics, your role becomes that of facilitator, supporting this engagement. Think about how students make sense of the number of intersections between two lines. When students work with multiple examples of pairs of lines, they can begin to see patterns and make conjectures in cases where there is zero, one, or an infinite number of intersections. How do students use the structure of the equations of these lines to help them determine these relationships? Students will begin to see how the slope and y-intercept of lines aid in determining the number of intersections in flexible ways. The development of this type of thinking supports students expressing Mathematical Practice 7, "Look for and make use of structure." How should teachers facilitate this sort of discussion with students so that the teachers are not doing all the telling (and thinking)? It requires instruction that acknowledges the value of students talking about mathematics and using mathematics to communicate their ideas.

When the mathematics content and the Mathematical Practices are addressed in tandem, students have the best opportunity to develop clarity about mathematical reasoning and what it means to do mathematics successfully.

Emphasizing the TQE Process

As part of the professional development material in this book, we include videos of high school classroom episodes (and occasionally grades prior to high school) in which students explore rich mathematical tasks. Classroom videos from grades before high school provide the opportunity to highlight the importance of prerequisite concepts and skills. For example, one of the included videos shows middle-grades students interpreting the growth of beams of a bridge, which demonstrates how unpacking and representing patterns in the middle grades connect to linking representations in high school. In presenting these videos, we emphasize three key aspects of the teacher's role—(1) tasks, (2) questioning, and (3) evidence—which make up what we call the *TQE process*.

Our emphasis on the TQE process helps define a classroom that develops mathematics as a focused, coherent, and rigorous subject. Thus, we uphold the following tenets.

- **Teachers with a deep understanding of the content they teach select *tasks* that provide students with the opportunity to engage in practices that support learning concepts before procedures; they know that for deep learning to take place, students need to understand the procedures they use:** Students who engage in mathematical tasks are also engaged in learning mathematics with understanding. Consider students who are making sense of transformations of quadratic functions. Students should explore the impact of different transformations on quadratic functions and begin to make conjectures about how these transformations are represented algebraically and graphically, as well as how they link to different contexts. Teachers know that this is important for later work in both algebra and geometry, as students will be working with transforming other types of functions, including trigonometric functions, and other representations, such as graphs of conic sections. These tasks should provide the impetus for students to make sense of transformations from both an algebraic and a geometric perspective. Students need tasks that allow them to build on what they already know and develop their own ideas and to talk with their peers about their thinking.
- **Teachers who have a deep understanding of the content they teach facilitate targeted and productive *questioning* strategies because they have a clear sense of how the content progresses within and across grades:** Using the idea of transformations, teachers know that it is important to build on students' thinking and guide them to make and test their own conjectures. Teachers plan learning opportunities for students to use their own strategies to overcome barriers, and they have set strategies in place to support students in learning from the barriers as they are encountered. If students struggle with translations, teachers ask questions such as "What do you notice about how the graph changes when you add four to the y-intercept? What happens when you subtract four?" to draw students' attention to how the graph changes with each vertical translation. Teachers need to be guided by student responses and ask supportive questions that maintain early focus on student-developed strategies.
- **Teachers who have a deep understanding of the content they teach use *evidence* gained from the formative assessment process to help them know where to linger in developing students' coherent understanding of mathematics:** Continuing with the example of understanding transformations, teachers who have a deep understanding of the mathematics they teach know that it is important for students to connect multiple representations of transformations to make sense of how each transformation is represented algebraically, graphically, and in the context of real-world situations. Students may use graph paper, transparent paper, graphing calculators, or online applets and programs to help them make sense of the transformation, and teachers will look for evidence of how students are linking representations together to measure students' understanding of the concept. The evidence should include how students are making sense of the impact of different transformations in all representations, rather than using procedures that might not be efficient, or worse, might be poorly understood.

Throughout the book and the accompanying classroom videos, we share elements of the TQE process to help you as both a learner and teacher of mathematics. In addition, we ask that you try to answer three targeted questions as you watch each video. These questions are as follows:

1. How does the teacher prompt the students to make sense of the problem?
2. How do the students engage in the task; what tools or strategies are the students using to model the task?
3. How does the teacher use questioning to engage students in thinking about their thought processes?

Next, we describe the structure of the book to help guide your reading.

The Structure of Making Sense

To address the mathematical content taught at the high school level, we had to make choices regarding what would be included within these topics and what would fall beyond the scope of this book. We decided to focus on the topics that connect algebra, geometry, and statistics together, not on particular courses. As such, some concepts are not included. We acknowledge that there may be future opportunities to make connections in these topics that we have not included in this book, and we look forward to continued work to connect important ideas in mathematics together.

To show ways in which algebra, geometry, and statistics interrelate, our chapter breakdown is as follows. Chapter 1 examines equations and functions. Chapter 2 then explores the structure of equations. Chapter 3 takes a look at geometry, and chapter 4 addresses types of functions. Chapter 5 examines modeling with functions, and chapter 6 closes with a focus on statistics and probability. These topics represent the big ideas for the high school grades. Each chapter concludes with a series of questions to prompt reflection on the topic under discussion.

We end the book with an epilogue featuring next steps to help you and your team make sense of mathematics for teaching and implement this important work in your school or district.

To further break down each overarching topic, each chapter shares a common structure: The Challenge, The Progression, The Mathematics, The Classroom, and The Response.

The Challenge

Each chapter begins with an opportunity for you to engage in an initial task connected to the chapter's big idea. We call this section The Challenge because this task might challenge your thinking, or perhaps the explanation of the task's solution will challenge how you think about using the problem with students. We encourage you to stop and engage with the task before reading further—to actually *do* the task. Throughout the book, we alert you to the need to stop and do tasks with a *do now* symbol (see figure I.4, page 10).

Figure I.4: ***Do now*** **symbol.**

The presentation of mathematical ideas in this book may be different from how you learned mathematics. Consider being asked to justify each step of the construction of an angle bisector. This task may test your understanding of constructions if it is not something that you have been asked to do before. The next step might be to ask you to construct a formal two-column proof that aligns to your construction, linking together different thinking processes in geometry, something you may or may not have considered in your own mathematics learning. Tasks in these sections focus on reasoning and sense making, since the rules of mathematics are developed through connections to earlier mathematical experiences rather than through procedures presented without meaning.

Since one purpose of this book is to engage you as a learner *and* teacher of mathematics, the tasks we ask you to explore support this goal. As a student of mathematics, you will consider how you learn mathematics. As a teacher of mathematics, you will explore how this newly found understanding could be the impetus for making sense of mathematics for teaching.

The Progression

Mathematics content knowledge alone is not enough. According to NCSM (2014), teachers must also "*understand how to best sequence, connect, and situate the content they are expected to teach within learning progressions*" (p. 24, emphasis in original). This means teachers need to know both the mathematics embedded in their course and the mathematics that comes before and after the course they teach—how the mathematics progresses over time.

Thus, each chapter highlights a progression of learning for a big idea. These progressions identify how learning develops over multiple years and highlight the importance of making sense of each building block along the way. The sequences defined by these progressions help the learner—and the teacher—make sense of the big idea in question. Understanding how content progresses provides avenues for supporting both the learner who struggles and the learner who needs enrichment.

Our placement of topics within learning progressions was informed by the Common Core State Standards for mathematics (NGA & CCSSO, 2010). However, our discussion of how the mathematics is developed within the progressions was not limited by this interpretation. We do not refer to specific content standards from the Common Core in an effort to expand the discussion to include *all* rigorous mathematics standards including those found outside of the United States. Note that because learning progressions develop over time, there will be occasions when this book addresses topics that reach back into the middle grades or forward into more advanced topics.

The Mathematics

There is much talk about rigor in instruction these days, but what does *rigor* mean in the context of teaching and learning mathematics? A misnomer is that rigor means *hard* or *difficult*. Rather, *rigor* refers to the need to incorporate all forms of thinking about mathematics—including concepts, procedures, the language of mathematics, and applications—in the teaching and learning process. However, this raises several questions. What does this actually look like in instruction? What teacher actions expose all students to rigor? How do you balance rigor reflected by reasoning and sense making with the ability to create generalized rules for finding the domain and range for a function or the maximum and minimum values of a function?

Problems in a rigorous classroom build students' deep understanding of mathematics. Consider how understanding function characteristics—such as domain, range, and end behavior—is developed using different types of functions. You know how to represent different characteristics of functions; you probably know them without needing to think. Yet what path do teachers use to bring students toward this level? All too often it begins and ends with memorization. Some of our students are successful memorizing the steps and procedures. Some are not. Is this an example of teaching with rigor? The answer is no. Rigor involves making sense of these characteristics based on the properties of functions.

Importantly, rigor includes making sense of the strategies involved in understanding the characteristics of functions. Sense making should precede a focus on memorization, rules, and procedures. This is where teaching might get uncomfortable. You may not have been taught the strategies as a student or see the connections. Thus, within The Mathematics section in each chapter, you will unpack the big ideas so the mathematics is explicitly connected to ways of making this knowledge accessible and rigorous for your students. Exploring mathematics in this way develops a deep personal understanding of mathematics. As you develop this understanding, procedures and algorithms will make sense, and you will be able to explain and justify them (NCTM, 2014). Through this process, we will support you in developing the knowledge to promote students' procedural fluency, which is defined as "skill in carrying out procedures flexibly, accurately, efficiently, and appropriately" (Kilpatrick, Swafford, & Findell, 2001, p. 116).

Many of the tasks throughout this book can be characterized as high-cognitive-demand tasks, which are "tasks that require students to think conceptually and that stimulate students to make connections that lead to a different set of opportunities for student thinking" (Stein & Smith, 1998, p. 269). Consideration for the understanding, creation, selection, and implementation of high-cognitive-demand tasks is vital for effective mathematics learning (Dixon, Adams, & Nolan, 2015). While including tasks that are high and low cognitive demand support a balance of conceptual understanding and procedural fluency, the cognitive demand of tasks often declines during instruction when the cognitive complexity of the task is not maintained (Kisa & Stein, 2015). How will you maintain the challenge of tasks through your actions? In this section and throughout the book, we provide you with excerpts from mathematics lessons so you can build a shared understanding of what mathematics instruction can look like in high school courses. In addition, we emphasize one or more of the best-aligned Mathematical Practices that supports the learning of the relevant mathematics content. This is important regardless of whether you are teaching using the Common Core State Standards for mathematics. It describes effective teaching and learning of any mathematical content.

The Classroom

In what ways do you support student learning in the mathematics classroom? For instance, do you encourage student discourse? Doing so allows you to consider what students are talking about and how you will respond to their talk. Your approach to student talk—and many other classroom aspects—helps determine the type of classroom learning community you and your students develop together. This classroom learning community is critical to the development of students' deep mathematical understanding.

The Classroom sections provide videos and extensive descriptions of what happens in engaging mathematics lessons. In order to assist you in thinking about how classrooms that develop mathematical understanding should look, each chapter includes two video episodes. These short videos show students exploring one task from the big idea of that chapter and one task from a related big idea. When you see the play button, please stop and watch the video. Included with the icon are the accompanying web address and a Quick Response (QR) code for you to access the video.

www.solution-tree.com/Interpreting_Graphs_to_Create_Context

We consider the accompanying videos to be a further investment in our effort to support the teaching and learning of mathematics. You will have the opportunity to see the topics we write about in the book in action. For instance, when we discuss how to massage a task so that it engages students, elicits student talk, and uncovers students' errors, we then follow up by capturing the essence of these actions in real classrooms. This modeling of good mathematics teaching provides opportunities for teachers to discuss what is happening in their own classrooms with the same mathematics content. It is not always possible for teachers to leave the classroom to observe a fellow teacher engaged in mathematics instruction. Thus, these videos help fill this gap and also provide a context for teachers to try an approach to mathematics teaching as modeled in the videos.

You must also consider the classroom expectations set for students. For instance, what are the rules for students answering questions? These rules should be established and made explicit for students so they know what you expect of them when they work on tasks. Also, how do students work together on tasks? Students can benefit greatly from collaborative experiences in mathematics, but they need to know how to best collaborate with each other. In many instances, the student who is confident and right most of the time does most of the talking. Thus, helping students monitor and regulate their discourse is valuable for the mathematics learning experience. As illustrated in the classroom videos, we encourage three classroom norms for every mathematics classroom.

1. Explain and justify solutions.
2. Make sense of each other's solutions.
3. Say when you don't understand or when you don't agree.

The classroom norms need to support the active thinking of students rather than solely relying on the thinking of teachers, as is so often the case in teacher-centered classrooms. This point is true when students work on their own as well as when students work together. You should always provide students the opportunity to share their strategies and make sense of the thinking of other students in order to be sure they understand mathematics with depth.

The structure of the classroom needs to support the thinking and learning of the students, and different tasks may require different structures. Some tasks may include questions that help students make connections; other tasks may not need such support. One model that is often discussed in many schools and districts is the gradual release of responsibility, commonly described as the teaching practice where the teacher models ("I do"), then the class practices together ("We do"), and finally the students practice independently ("You do") (Fisher & Frey, 2003). Although this method is appropriate at times, we advocate for methods that include focus on the *students* making sense of and reasoning with mathematics. An alternative approach for mathematics is what we call *layers of facilitation*.

1. *I facilitate the whole class* to engage in meaningful tasks through questioning.
2. *I facilitate small groups* to extend the learning initiated in the whole-group setting.
3. *I facilitate individuals* to provide evidence of their understanding of the learning goal.

This change in teacher role focuses on the teacher as a facilitator of knowledge acquisition rather than as a transmitter of knowledge. As you read the text and as you watch the videos, you will notice our focus is largely on implementing layers of facilitation, but it is important to recognize that there are some topics and mathematical content that need to be taught following the gradual release of responsibility model. It is important for you, as the teacher, to determine when to apply this model. Keep in mind that students will benefit from opportunities to have a more participatory role during instruction whenever possible.

The Response

How do you respond when students struggle? What do you do when students express misconceptions? It is important to use student errors as springboards for learning; the errors and the gaps in prerequisite knowledge that lead to those errors inform your everyday instruction as well as your response to intervention (RTI) process within a Multi-Tiered System of Supports (MTSS).

Again, consider how students might engage with the two approaches to solving $x^2 - 5x + 4 = 2(x - 1)$. Perhaps they would see the connection between the algebraic approach and the graphical approach. How would you extend students' thinking when this occurs? Think about what questions you should ask in order to shed light on students' thinking processes, both when they are making appropriate connections and when they struggle. When students mistake the solutions as the y-values of the intersections of the graphs instead of the x-values, how will you help them examine their thinking and make the correct selection without simply giving them the correct answer? One approach would be to ask questions to support the efforts of your students and encourage them to think of errors as expected and beneficial in order to learn mathematics with depth. This is an area where your own depth of mathematical understanding is critical to help your students advance their thinking.

It is essential to explore your students' reasoning and sense making and break down that thinking in order to rebuild their understanding of mathematics. Effective teachers understand what models and strategies best support their students in ways that allow them to connect their current thinking with the learning goal of the task. You can then use the evidence of the level of students' understanding you gather to inform your response both during and after instruction.

Now that you understand what is to come in the following pages, here's how we suggest you approach the book.

How to Use This Book

Collaborative teams of teachers can use this book to explore mathematics content and engage in discussions about teacher actions that will help bring mathematics to life for students. In fact, the entire TQE process is best accomplished by a collaborative team that works together to address the four critical questions—the guiding force of the professional learning community (PLC) culture (DuFour, DuFour, Eaker, & Many, 2010):

1. What do we want students to learn and be able to do?
2. How will we know if they know it?
3. How will we respond if they don't know it?
4. How will we respond if they do know it?

This book is an optimal tool for your collaborative team in a PLC culture. Although it is grade-band specific, it also provides support for vertical (across grades and courses) discussions and planning. There are many topics in this book that can be addressed in your course team or in your vertical team, including the Mathematical Practices. While the Mathematical Practices are not course specific, how students engage with them can be expected to vary from course to course, and you can benefit from course-specific as well as vertical discussions about them.

Our expectation is that individuals or collaborative teams will be able to use this book by reading the chapters in order. Within each chapter, we help you develop clarity about the mathematical content and its progression. Teachers often have questions about the sequence of mathematics content in mathematics texts and other resources—sometimes the sequence is aligned with authentic progressions and sometimes not. Thus, you can use this book as a resource to understand how background and underlying knowledge of mathematics support further understanding and how to best align mathematics curriculum with the progression of the content.

We hope this book and the accompanying videos will be your go-to sources for a deep dive into relevant mathematics content, effective pedagogical actions, appropriate classroom norms, meaningful assessment, and collaborative teacher team efforts. Our goal is for this resource to connect the good work of mathematics teaching that you are already facilitating with the goals of improving mathematics teaching that you aspire to attain.

CHAPTER 1

Equations and Functions

What is the purpose of teaching functions in secondary mathematics? The concept of functions is one of the most important concepts in mathematics; it enables students to make connections within and across mathematical topics. Students need to understand patterns and the link to functional relationships, learning that mathematics is more than just manipulating algebraic expressions. As students use functions to represent relationships in multiple ways, they begin to recognize mathematical ideas in real-world situations and form notions of mathematical modeling (see chapters 5 and 6 for discussions on modeling). In order to emphasize the importance of exploring and connecting different types of functional relationships, different function types, such as linear and quadratic functions, and systems of linear equations are explored through real-world situations in this chapter and in chapter 2. Connections are made among verbal, numerical, graphical, and algebraic representations of mathematical concepts to illustrate the role and significance of multiple representations in understanding functions.

The main goal of this chapter is to develop deep understandings of equations and to examine the underlying equivalency of representations by modeling and predicting real-world phenomena. This is accomplished by providing an overview of the progression of early algebra learning and the notion of expressions, equations, and variables, then defining and comparing linear and nonlinear functions in multiple representations.

The Challenge

The initial task in this chapter (see figure 1.1, page 16) begins the process by providing a contextual application of linear functions. Complete this task before reading any further.

How did you construct your story? What is the importance of the *y*-intercept in the given context? What is the importance of the intersection point? How did you determine each person's speed? Which person was running faster? Does your story include who won the race?

The race between father and son is represented by linear graphs. Although one line is steeper than the other line, both distances are increasing at constant rates. In other words, the father and son are both running at constant speeds, even though these are not the same speed. In exploring the task, it is important to reflect on what each of the linear graphs represents, why the speeds are constant, and the importance and the mathematical consequences of the speeds being different. What is the meaning of the slope in the context you wrote? Perhaps you concluded that the slope of each line represents the speed of each person and that speed is a rate of change that relates distance and time. How is the father's change in distance related to the time he has run? How does this rate of change compare to the son's change in distance related to the time he has run? Table 1.1 (page 16) examines the distance run at intervals to help determine the speed of each runner, or the slope of the graph.

DO NOW

The figure below shows the relationship of distance and time for a father (F) and son (S) during a 100-meter race. Write a story that matches the graph. Describe each stage of the race. In particular, describe and explain what is occurring at times A, B, and C, as well as the intervals in between.

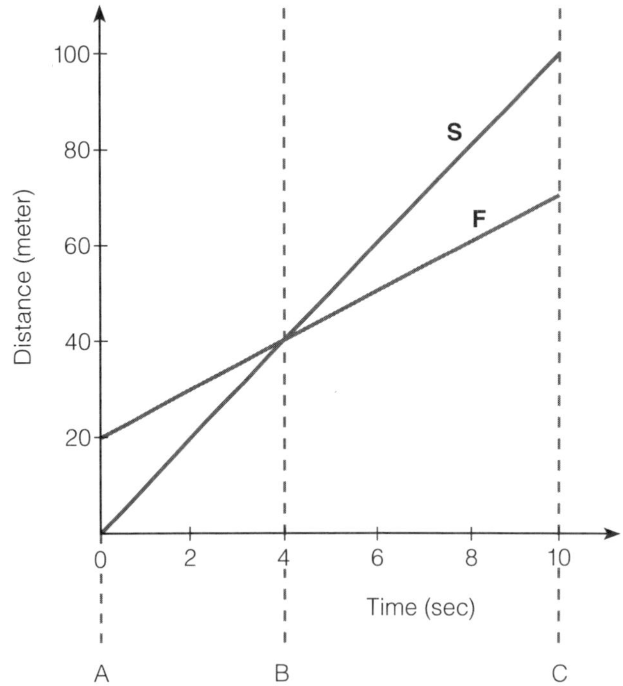

Figure 1.1: A 100-meter race between a father and son.

Table 1.1: Distance Run Over Different Intervals

	Father	Son
Distance run between A and B	20 meters	40 meters
Distance run between B and C	30 meters	60 meters

How much time elapses between times A and B? How does this help you determine the speeds of the father and the son? The father runs 20 meters in 4 seconds, or 5 meters per second. In comparison, the son runs 40 meters in 4 seconds, or 10 meters per second. Is the speed between times B and C the same as it is between times A and B? For the father, the speed is 5 meters per second (30 meters divided by 6 seconds); for the son, the speed is 10 meters per second (60 meters divided by 6 seconds); therefore, the speeds are the same.

Re-examine the story you wrote for this task. What, if anything, would you change in your story after examining the speed in this way? Determining how to compare the speeds of the father and the son is an important consideration for how you wrote your story to describe the graph. As you examine the story, ask yourself how you took into account the y-intercept of each line. What do those points mean within the context of the story? As the race begins, the father starts from 20 meters, whereas the son is at the starting point of the race. This means the father has a head start when the race starts. What is the significance of time B? It represents the time it takes for the father and son to meet in the race assuming that they are on the same track, running in the same direction, and running to the same finish line. After 4 seconds, they are both 40 meters from where the son started the race. What happens then? You know from your analysis of the slope that the running speeds don't change, so why does the son's graph appear to be "above" the father's graph after this point? This illustrates that time B is where the son catches up to his father and then passes him. Between times A and B, the son is behind the father in the race. After time B, the son is ahead of the father in the race. For your story to completely describe the graph, you should have included where the father and the son met and what happened before and after they met.

Did your story indicate who won the race? At what time in the race does it become clear that the son will win? Four seconds after the race begins, the son catches up to his father. This occurs at 40 meters, or time B. The son reaches the finish line in 10 seconds and wins the race by 30 meters, at time C. What if the son does not want to win the race by such a great margin? He can decide to give his father a greater head start. This is an opportunity to build on what is in your story and extend your thinking about this context. What would happen if, in addition to a 20-meter head start, the son waits 2 seconds before beginning his race and the running speeds are maintained from the original task? How would this change the graph of the son's line? The slope would remain the same, but the y-intercept would be –20 rather than 0 (if you represent this situation with a linear graph). Would both lines change? No, only the son's line would change. How would it change the story? The son would actually win this race as well, completing the race in 12 seconds. How long would he need to wait in order to lose the race if he gives his father the same head start? As it would take the father an additional 6 seconds to reach 100 meters, the son would need to wait more than 6 seconds in order to lose the race.

Making sense of graphs by creating contexts to match changes in graphs becomes a critical component of reasoning algebraically, understanding the relationships between contexts and graphs with depth, and using equations, expressions, and functions in meaningful ways. It also provides an excellent opportunity to engage in Mathematical Practice 2, "Reason abstractly and quantitatively."

The Progression

As you examine algebraic concepts of equations and functions, it is important to consider where the development of these concepts begins. Ideas for the notion of variable and algebraic expressions are introduced in grade 6 and continue through grade 8 and into high school. In grade 7, students generate equivalent expressions and construct equations to solve real-world problems using equations. In grade 8, the focus shifts to defining and representing functions with multiple representations. (For further discussion of these ideas, see chapter 3 of *Making Sense of Mathematics for Teaching Grades 6–8* [Nolan, Dixon, Roy, & Andreasen, 2016].) This leads to the following progression of equations and functions in high school.

- Construct equations in two variables.
- Analyze functions in multiple representations.
- Use functions to model real-world situations.
- Define and compare linear and nonlinear relationships in multiple representations.
- Connect domain, range, and inverse of a function to its graph.

Grade 7

In grade 7, students use variables to construct expressions and equations to solve real-world problems. Students learn to apply properties of operations to generate equivalent expressions and explore proportional relationships using equations. Students are expected to describe and explain relationships and represent them with tables, graphs, and equations.

Grade 8

In grade 8, students learn to define and model relationships between quantities. They also compare linear and nonlinear relationships using multiple representations. As students create equations in two variables and model relationships between quantities with multiple representations, they begin to develop a dynamic view of functions and prepare themselves for work with complex functions in later grades.

High School

In high school, students extend the work with linear equations and functions to construct functions between two quantities and model real-world situations. When presented with a linear relationship between quantities, students learn to construct an equation in two variables, compute the rate of change between the two variables, and interpret the rate of change in real-world situations. As they explore functional relationships verbally, numerically, graphically, and algebraically, they develop flexibility in translating among these different representations. They define and compare linear and nonlinear relationships, and this enables them to learn to distinguish among real-world situations regarding those that can be modeled with linear functions and those that cannot. Students in grade 8 become familiar with the concept of functions, and concepts such as domain, range, and the inverse of a function are introduced in high school coursework.

The Mathematics

Equations and functions can be used to model real-world situations and mathematical relationships. High school students must be able to analyze functions with multiple representations and connect the domain and range of a function to its graph. This is best accomplished by using a variety of situations so that students are able to establish meaningful connections among representations and consider reasons for the use of a particular representation.

Analyzing Functions With Multiple Representations

Consider the task in figure 1.2. Take a moment to make sense of this task before continuing.

DO NOW

> The length of a rectangle is twice its width. Create an equation that relates the perimeter of the rectangle to the width of the rectangle. Determine the perimeter for rectangles of various widths. Create several representations for describing the relationship.

Figure 1.2: Perimeter of a rectangle.

What equation did you create for the perimeter of the rectangle? Since the length is twice the width, if you use x for the width, the length would be $2x$. How do you represent the perimeter of a rectangle? In this case, the perimeter can be represented as $x + 2x + x + 2x = 2(x + 2x) = P$. This can be simplified to be $P = 2(3x) = 6x$. What does this indicate about the relationship between the width of the rectangle and the perimeter? As the perimeter of this rectangle is a constant multiple of the width, there is a proportional relationship between the perimeter of the rectangle and the measure of its width. The perimeter of the rectangle is therefore dependent directly on the measure of its width.

In what other ways did you represent the relationship? You may have chosen to use a set of points, a table, a mapping, or a graph. How can you represent various widths and perimeters using a set of points? Each pair of width and corresponding perimeter can be represented as an ordered pair using set notation, such as {(1, 6), (2, 12), (3, 18)}. In this case, the first element in the ordered pair represents the width of the rectangle and the second element is the corresponding perimeter. As the order in which the pair of elements is listed is important—because switching the elements represents a different relationship—this manner of expressing pairs of elements is referred to as *ordered* pairs. A set of ordered pairs is called a *relation*. The first element in the ordered pair is considered the independent variable; the second element is considered the dependent variable. Why in this context is the independent variable the width of the rectangle and the dependent variable the perimeter? This is because the perimeter depends on a given width.

Using set notation represents one method to express how elements of one set are related to elements of another set; however, other visual and numerical representations can be effective when exploring mathematical relations between quantities. In this case, if you consider two sets with a finite number of elements in each, you can create a table of values in which elements of each set are placed in a different row (or column). You can also represent the same relation with an arrow diagram connecting the elements of the sets in a mapping as shown in figure 1.3.

Measure of Width	1	2	3
Perimeter	6	12	18

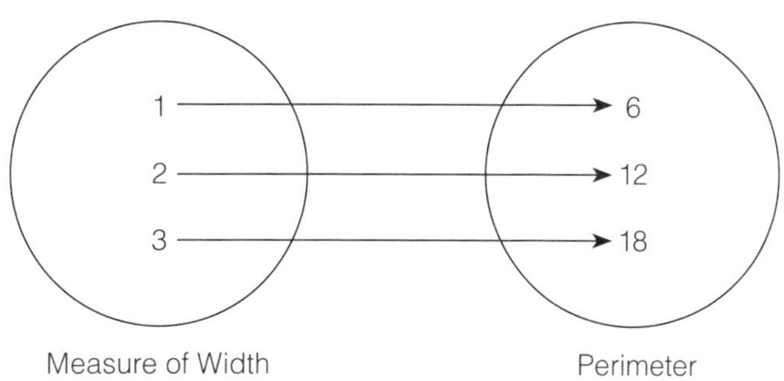

Figure 1.3: Representation of the perimeter of a rectangle using a table and a mapping.

In the middle grades, students make sense of ratio and proportional reasoning using double number lines. (See chapter 2 of *Making Sense of Mathematics for Teaching Grades 6–8*.) This can be connected to graphical representations of these mappings as well. Since both sets consist of real numbers, you can consider the elements of these sets as values on a number line. If you then show how the elements of the two sets are related to one another, you create a double number line, which, in this case, relates the width of the rectangle and its perimeter. The diagrams in figure 1.4 show two ways that a double number line can be used to relate the width and perimeter. Using the double number line provides a way to make sense of how the perimeter is changing given the measure of the width.

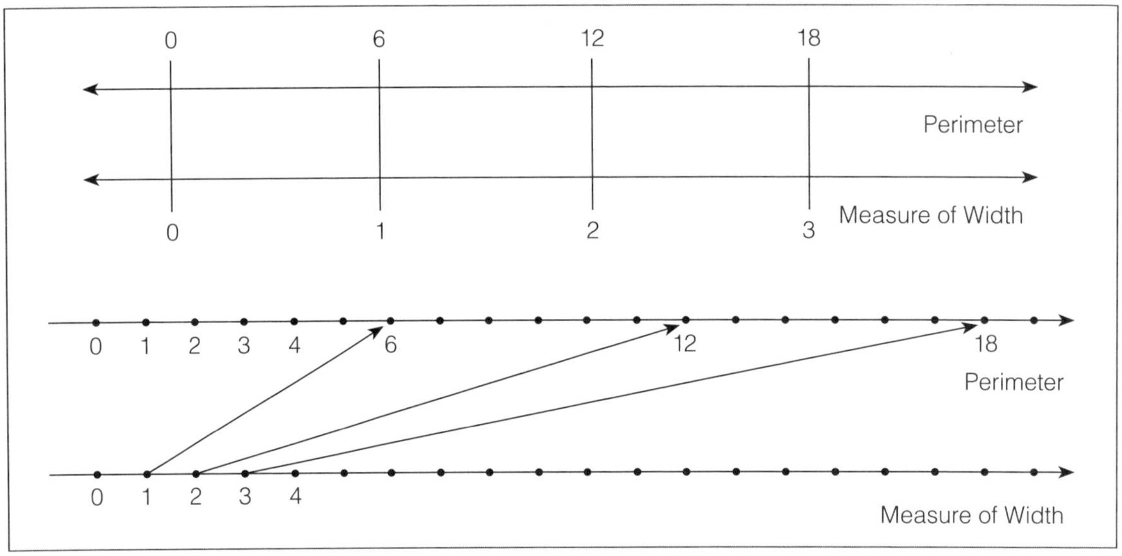

Figure 1.4: Two representations with double number line.

What other representations could you use to show this relationship? Instead of relating elements of each ordered pair with arrows on two number lines that are parallel to one another, each ordered pair can be represented as a single point using two number lines that are perpendicular to one another (the Cartesian coordinate plane).

The bottom diagram in figure 1.4 demonstrates how the same scale for two different number lines would be represented using a double number line. If you imagine placing the number line corresponding to the perimeter on top of the number line corresponding to the width and rotating only the top number line of this double number line 90° counterclockwise, this would create a Cartesian coordinate system in which the horizontal axis is the width (the independent variable) and the vertical axis is the perimeter (the dependent variable). The Cartesian coordinate plane offers a powerful method of plotting ordered pairs and visually representing relations. This allows you to connect the double number line to a graphic representation of the proportional relationship (see figure 1.5).

The graph represents a proportional relationship, as the y-intercept is 0 and the rate of change is constant. For this proportional relationship, the rate of change, or slope, is 6 units, as the value of the dependent variable increases 6 for every unit increase of the independent variable.

A negative slope with a non-zero y-intercept is more complicated to map. Consider the double number line map of $y = -2x + 3$ in figure 1.6.

How does this double number line differ from the perimeter example of $y = 6x$? Note how the arrows that map the x-values to the y-values cross over one another, compared to the arrows in figure 1.4. How are positive and negative slopes represented differently on the Cartesian coordinate plane? Think

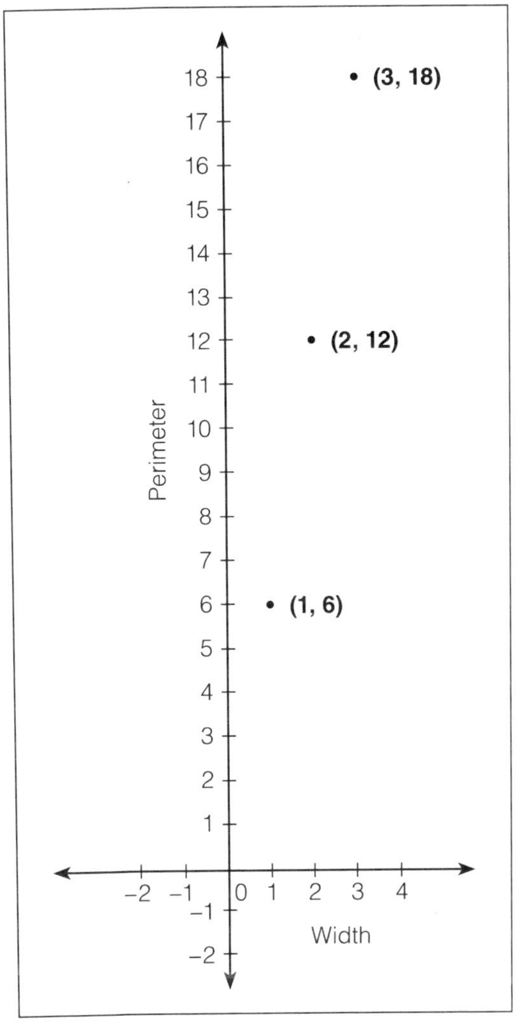

Figure 1.5: Double number lines as a Cartesian grid.

about how the representation will differ if you situate the *y* number line over the *x* number line (figure 1.6) and rotate the *y* number line 90° counterclockwise. The points will be in the first, second, and fourth quadrants of the coordinate grid (see figure 1.7, page 22). Exploring the connections among representations provides you the opportunity to engage in Mathematical Practice 7, "Look for and make use of structure." By focusing on how the representations are connected, you are making sense of how they display information.

Reconsider the perimeter problem (see figure 1.5). In this case, each of the ordered pairs reinforces that the perimeter is six times the measure of the width (or the output is six times the input). All of the elements of the set of measures for width (or inputs) that are part of the relation create the domain, and all elements of the perimeter (or outputs) that are part of the relation create the range. What values could be possible for the elements of the domain and the range? Could they be negative values? Fractional values? Considering this context, the domain and range should be positive rational numbers. In other words, because the width represents a measure of length, the value of *x* cannot be zero nor could it take on negative values. Should the graph be represented as discrete points or as a connected line? To represent this relation on the coordinate plane, a straight line should be used instead of discrete points because, according to this relation, between any two points, there is another point, so the line representing this relation is continuous (see figure 1.8, page 23).

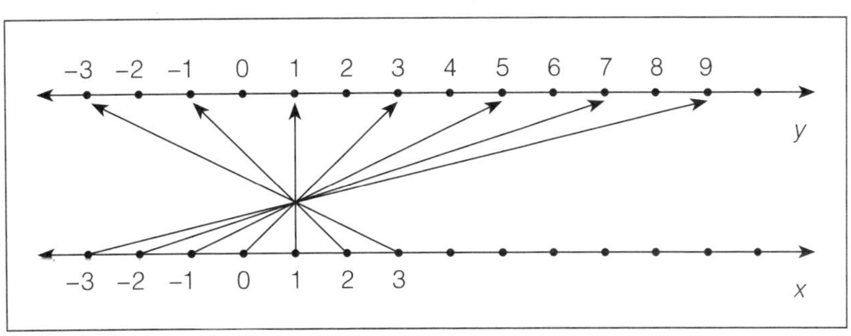

Figure 1.6: Representing the relationship of $y = -2x + 3$ with the double number line.

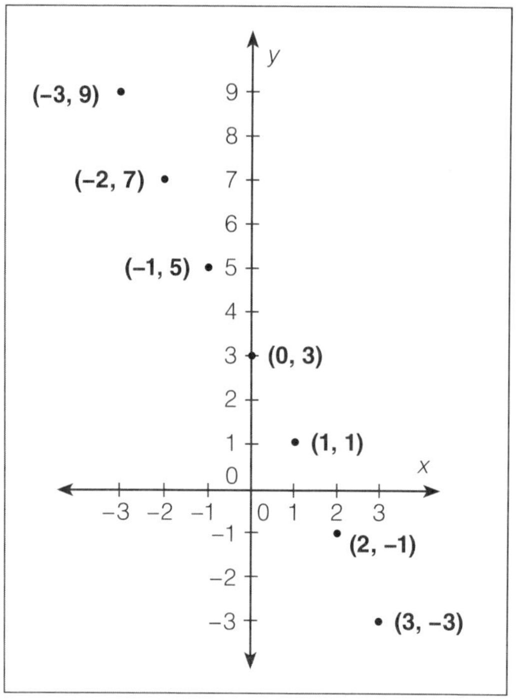

Figure 1.7: Representing the relationship of $y = -2x + 3$ with the Cartesian grid.

What is the rate of change of this graph? In this case, as the width changes by one unit, the perimeter changes by six units. The rate of change is described as the amount of change in the dependent variable (the y variable) in relation to the amount of change in the independent variable (the x variable); therefore, the rate of change is six units of perimeter per unit of width. The slope emphasizes the dependent relationship between the width and the perimeter. The rate of change is also represented by the slope of the line.

How might the situation be changed so that the relationship between the measure of the width and the perimeter has a greater rate of change? If, for example, the length is three times the width, then the perimeter is equal to eight times the width. Figure 1.9 provides a visual comparison of the two scenarios.

The graph of the perimeter of a rectangle in relation to its width creates a linear representation. Some relations are also considered *functions*. Is this relationship between the width of the given rectangle and its perimeter a function? A function is a relation in which each input value (a value of the independent variable) is mapped onto only one output value (a value of the dependent variable). Could you make two different rectangles having the same width that fit this context but have different perimeters? As each rectangle with a given width in this situation has one and only one perimeter, this linear relation is classified as a function. How can you determine whether a nonlinear relation can be a function? Consider the task in figure 1.10.

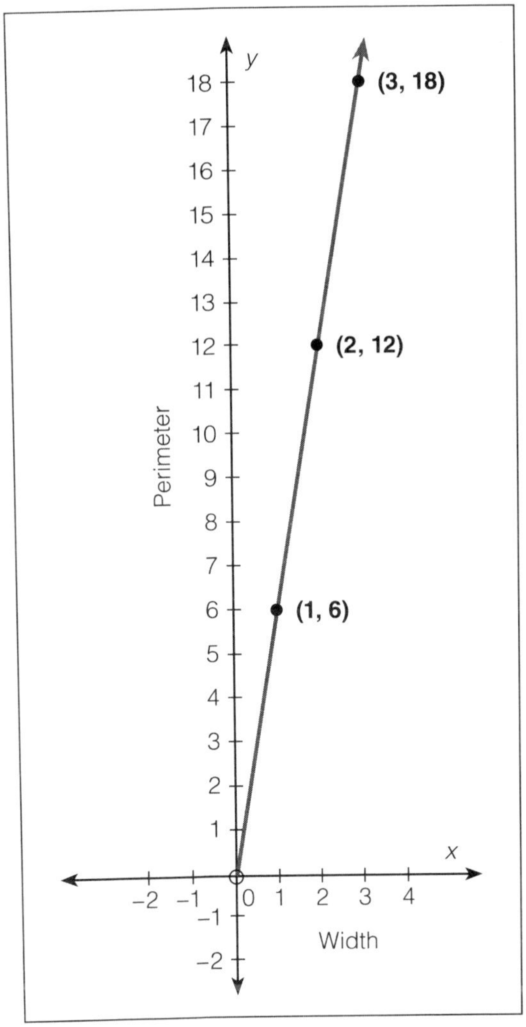

Figure 1.8: Continuous graph of the relation between perimeter and width.

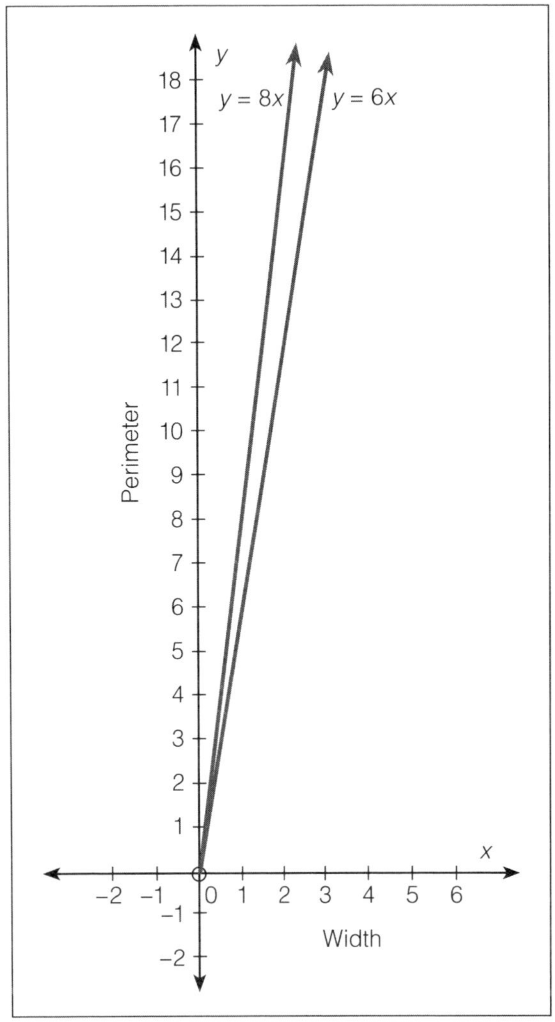

Figure 1.9: Comparing different rates of change.

DO NOW

The length of a rectangle is twice the width. Create an equation to relate the area of the rectangle to the width of the rectangle. Determine the area for rectangles of various widths.

Figure 1.10: Area of a rectangle task.

How did you create your equation? The area of the rectangle in relation to its width is a nonlinear function. If the length of the rectangle is twice its width, then the function for the area in terms of its width is $A = x(2x) = 2x^2$. The *y*-coordinate of every point on the graph of this relation is twice the square of the *x*-coordinate. For instance, the following points are on the graph of this area function: (1, 2), (2, 8), and (3, 18). The width of the rectangle in this situation has only one area value, so here again, the relation represents a function. How do you know this function is nonlinear? Students explore these types of functions and conjecture that the area function is not linear because these points do not create a straight line on the graph (see figure 1.11, page 24).

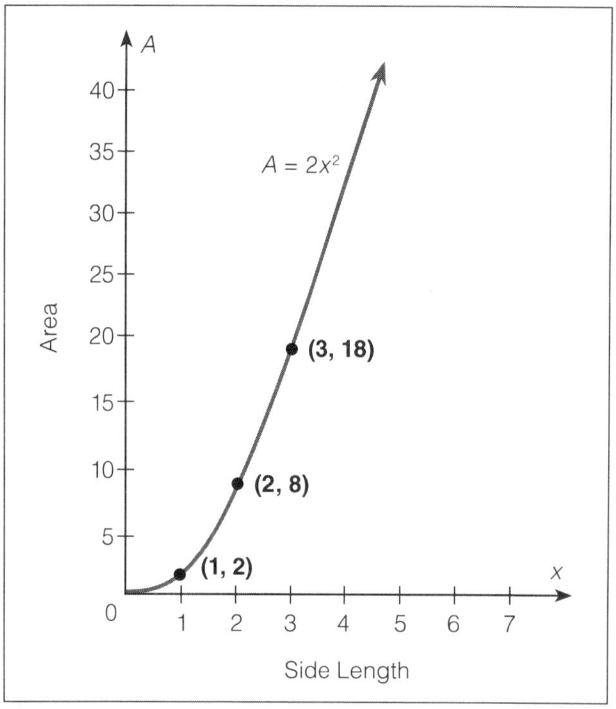

Figure 1.11: Area as a function of side length.

What would happen to this function if you changed the relationship to have the length equal to four times the width? The area would be modeled by the function $A = 4x^2$. Think about the impact of the change in the coefficient from 2 to 4. How does this change the relation? Consider figure 1.12.

How does the change in coefficient from 2 to 4 impact the different representations? As the coefficient of x^2 is increased, the values of the area increase as well. One unit change in the x-value causes a greater change in the y-value. In relation to the y-axis, the graph for $A = 4x^2$ is located "above" the graph of $A = 2x^2$. This corresponds with the values in the table increasing at a faster rate and the graph increasing in a steeper fashion. The graph reflects this change visually through a steeper rate of change, while the table shows this through a numerical representation of the different rate. As the rate of change is not constant, the functions are not linear functions. Visually, the graph shows that the rate has changed, and the table allows you to quantify this change precisely.

Connecting Domain and Range of a Function to Its Graph

In mathematics, functions are known as real (or real-valued) functions when their domain and range consist of real numbers. While a function might be a real function, this does not mean that its domain and range necessarily contain *all* of the real numbers. There might be restrictions on the domain and range of a real function. These restrictions could be generated by the function itself or arise due to the context to which it is connected. In determining the domain and range, you should determine what input and output values can and cannot occur for a given function.

Consider the father and son race in figure 1.1 (page 16). What restrictions exist for the domain as a result of the context? Can time be negative? Does the father or the son run farther than 100 meters based on the graph and the story? In this context, the domain for both functions is restricted to values between 0 and

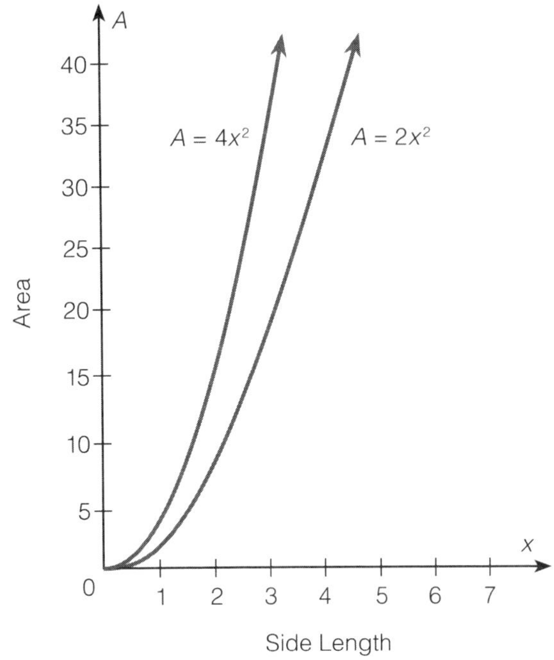

$A = 2x^2$

Side Length (x)	Area (A)
1	2
2	8
3	18

$A = 4x^2$

Side Length (x)	Area (A)
1	4
2	16
3	36

Figure 1.12: Area comparison with graph, equations, and tables.

10 seconds because the story ends when the son wins the race after ten seconds. The range is restricted to values between 0 and 100 meters. These restrictions are based on the context of the problem, not the functions themselves. When taken on their own, the mathematical representations of these functions appear to have a domain and range of all real numbers; however, when taking the context into consideration, this is no longer the case.

What might be other reasons for a particular value or values to be excluded from the domain or range of a function? What kinds of relations would have no restrictions on their domain or range? Consider the task consisting of relations presented verbally, graphically, and algebraically in figure 1.13 (page 26).

How did you determine the domain and range for each relation? How did you identify which values should be excluded from the domain? The first problem is a linear relation. All input and output values are possible in this relation. Therefore, you can say the domain and range of this relation consist of all real numbers. How can you determine whether this relation represents a function? What does this relation look like graphically? The relation $y = 5x$ describes a rule that applies to all points that lie on the graph of this relation; that is, if an ordered pair lies on its graph, then the y-coordinate is five times the x-coordinate.

DO NOW

Determine the domain and range of each relation, and determine if it is a function.

1. $y = 5x$
2. A child holds a ball 1 foot off the ground and tosses the ball straight up in the air. The ball reaches a height of 30 feet and then falls to the ground. This occurs over a 3-second period, with the ball reaching its peak at 1.5 seconds.
3.

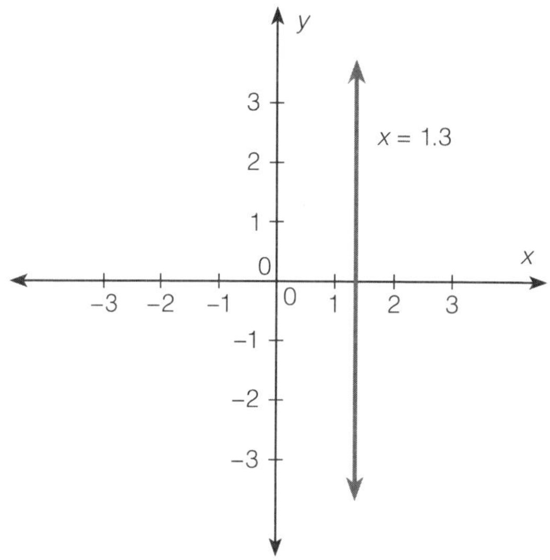

Figure 1.13: Determining domain and range.

The point (1, 5) is on the linear graph because the value of y is five times the value of x. This point can be plotted on the coordinate grid at the intersection of the set of points where $x = 1$ and the set of points where $y = 5$ (see figure 1.14).

Is the relation a function? Any intersection point between a vertical line and the linear graph on the coordinate plane indicates the existence of a y-value in the range corresponding to an x-value in the domain. If a vertical line intersects a graph at two or more points (meaning that the graph contains two or more points with the same x-coordinate), then an x-value in the domain corresponds to two or more y-values in the range and the relation is not a function. In these cases, the graph fails the *vertical line test* for determining if a relation is a function. Reconsider figure 1.14 to see if the relation $y = 5x$ is a function.

What would be the meaning of a *horizontal* line test? What does an intersection point or points between a horizontal line and the graph represent? An intersection point indicates the existence of an x-value in the domain corresponding to the y-value in the range represented by the horizontal line. If a horizontal line intersects a graph at two or more points (that is, the graph fails the horizontal line test), then a y-value in the range corresponds to two or more x-values in the domain. While this does not impact whether the relation is a function, it does indicate that the relation is not a one-to-one function. This will be an important consideration when you examine inverse relationships later in your reading.

Is it possible to draw a vertical or a horizontal line that intersects the graph of $y = 5x$ at two or more places? If you draw a vertical line and move it from left to right, you will see that the vertical line appears to intersect the linear graph in at most one point. If you draw a horizontal line and move it up and down,

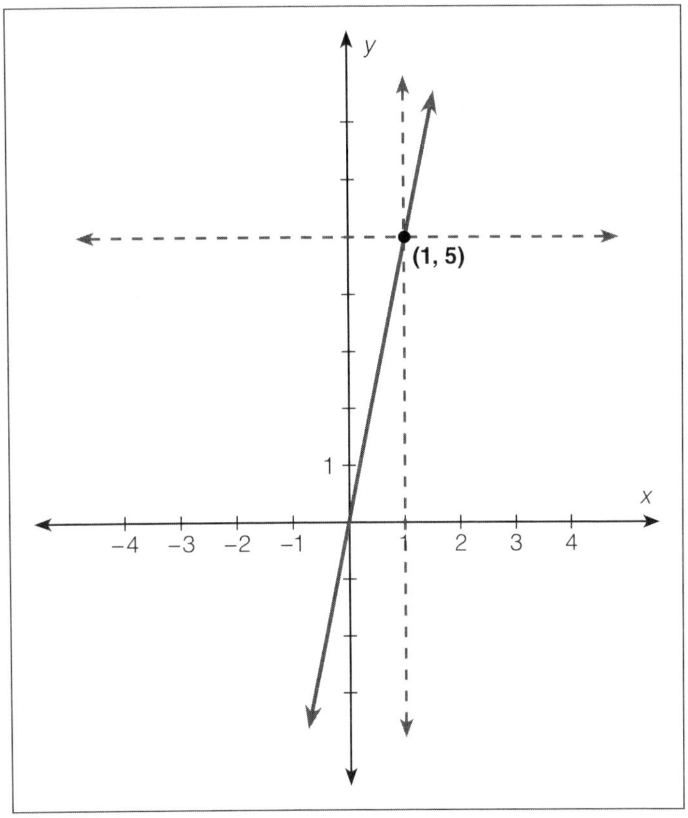

Figure 1.14: Graph of $y = 5x$.

it will also intersect the linear graph at exactly one point. Both of these conditions indicate that the relation $y = 5x$ is a one-to-one function. Often students may understand the identification of functions, nonfunctions, and one-to-one functions but fail to recognize them in a context. What do the vertical and horizontal line tests mean for a given context?

Consider the following distance-time context applied to problem 1 in figure 1.13: "Javier is traveling at the speed of 5 miles per hour. How many miles will he travel in 1, 2, 3, or x hours?" In context, the number of hours traveled would represent only one distance and any distance would be generated from only one time period. As previously discussed, the graph of this relation passes both the vertical and horizontal line tests; therefore, the graph represents a one-to-one function. In this distance-time graph, the vertical line test allows for the explanation of whether Javier can travel different distances at any given time. The horizontal line test asks if Javier can be at the same distance at different times. The context, represented by $f(x) = 5x$, supports a one-to-one function, but only for positive values in the domain and range since, in this context, negative time and negative distance do not make sense.

Now consider problem 2 from figure 1.13. What are the domain and range of this relation? The independent variable is time, and because of that, the domain cannot be negative. Since the entire ball toss lasts just 3 seconds, the domain is from 0 to 3, inclusive. The ball starts at 1 foot above the ground, so students often assume the range is from 1 to 30, the highest point attained by the toss. However, since the ball eventually lands on the ground, the range is actually 0 to 30, inclusive.

Is this context modeled by a function? For every x-value, there is only one y-value. As time passes, the ball is never in two places at once. Therefore, the context is modeled by a function. However, the function is not a one-to-one function because the ball is at the same height at two different times, both on its way up and on its way down, so there are instances in the relation when there are two x-values (time) for an individual y-value (height).

How do you determine the domain and range from a graph? Since the domain is associated with the independent variable, the x-axis provides information for the values in the domain. For the graph in problem 3 of figure 1.13 (page 26), the domain is the single value $x = 1.3$. The y-axis values of the line, which represent the values that the dependent variable takes on, are the values for the range. The range for problem 3 is all real numbers.

How do you determine if the graph in problem 3 from figure 1.13 is a function? What if the graph actually *is* a vertical line? Because there is only one domain value and more than one range value for that domain value, this graph does not represent a function, as the single value of 1.3 produces all of the range values of the graph. In order for a relation to be a function, any x-value cannot be mapped to more than one y-value. In this case, the same x-value is mapped to infinitely many y-values. Therefore, this cannot be a one-to-one function since it is not a function initially.

Why is it important to identify one-to-one functions? The answer is that only one-to-one functions have an inverse that is also a function. This creates more questions: What is an inverse function, and what does it mean in this context?

A function performs a rule on each x-value in the domain to produce a single y-value in the range. An inverse function operates in reverse. This is comparable to thinking about evaluating an expression. Consider the expression $2x + 5$. For a given x-value, how do you determine the value of the expression? In this case, you would first multiply by 2 and then add 5. If, however, you knew the value of the expression was 17, how would you find the value of x that produces that value? You would first subtract 5 and then divide by 2. This is comparable to a function and its inverse. The function defines what happens to an input to determine the output; the inverse defines what happens to the output to determine the necessary input. How are the graph of a function and its inverse related?

Return to the equation $y = 5x$. You can determine its inverse algebraically and use this new equation to draw the graph of the inverse function.

$$y = 5x$$

then consider how you would determine the value of y if given the x value,

$$x = \frac{1}{5}y$$

Placing this relationship into the functional notation

$f(x) = 5x$ has the inverse function $f^{-1}(x) = \frac{1}{5}x$

How are the graphs of a function and its inverse related? You can also use the graph of the function $y = 5x$ to draw the graph of its inverse. That is, when you graph the inverse function, you interchange the x-values and the y-values. This means that the graphs will be symmetric. You can reflect the graph of $y = 5x$ over the line $y = x$ to determine the graph of the inverse function. Figure 1.15 provides a visual representation of the result of this process.

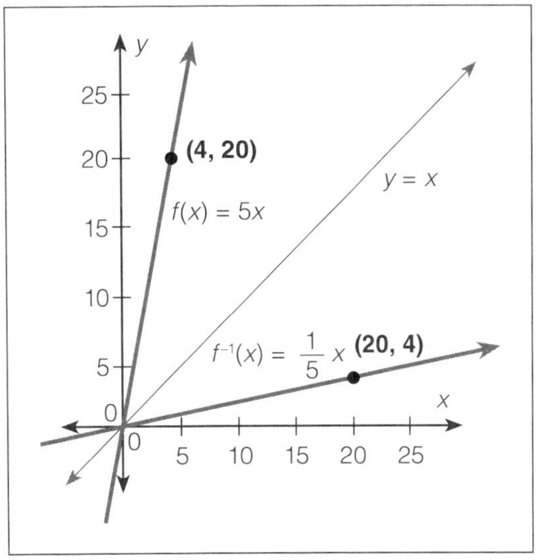

Figure 1.15: Graph of a function and its inverse.

Take an x-value and substitute that value into the function. For instance, if $x = 4$ in the function $f(x) = 5x$, then $f(4) = 5 \times 4 = 20$. As an ordered pair, this is written as (4, 20). Now take this y-value and substitute it into the inverse function $f^{-1}(20) = \frac{1}{5}(20) = 4$. As an ordered pair, this is written as (20, 4). You see that the x- and y-coordinates switch places. In general, since the function and its inverse work in opposite directions, the set of elements comprising the domain and the set of elements comprising the range switch places. As a result, any point (x, y) on the graph of the function becomes the point (y, x) on the graph of its inverse. This is true for any point on the graph of a function and its inverse.

How can the inverse function be applied to the relationship between distance and time? Consider the task in figure 1.16.

Create a scenario that *reverses* the role of time and distance from the function of $f(x) = 5x$ and the context "Javier is traveling at the speed of 5 miles per hour. How many miles will he travel in 1, 2, 3, or x hours?" Explain the role of the independent and dependent variables in your new scenario.

Figure 1.16: Exploring the role of the independent and dependent variables in context.

In the original scenario, the goal was to estimate how far Javier would travel given a time frame. In your new scenario, you want to determine the amount of time that it takes to travel a particular distance. How can the ideas of inverse functions be applied to the relationship in the context of time and distance? You may have written, "Javier is traveling at the speed of 5 miles per hour. How long will it take him to travel 2, 5, 10, or x miles?" When people travel distances for a specific purpose, such as going to work or visiting tourist destinations, time plays a key role. You may want to know, given different distances, how long it may take you to travel to each of those destinations. Perhaps you are considering homes that are different distances from your place of employment; if you travel at the same speed, which one would be the shorter (time) commute? The dependent variable is time, and the independent variable is distance. The inverse function of the original time and distance relationship would be used to model this scenario in order to determine the amount of time it would take to get to your destination.

Analyzing problems in context helps you and your students make sense of the definition and characteristics of functions. Exploring how domain and range help reveal whether a relation is a function, and whether the function is one-to-one, is an important step in understanding functions. Understanding these characteristics also provides the foundation for developing meaning for inverse functions.

The Classroom

Now that you have explored your own thinking about functions, you will explore what this looks like in the classroom. The following videos provide some important insight into student thinking about equations and function relationships. In the first video, you will watch grade 8 students exploring a task to make sense of equations of lines and their respective graphs, particularly determining when lines are parallel or intersecting. This task prepares students to consider the characteristics of lines that will support understanding systems of equations with depth (this is addressed in chapter 2). Take a moment to watch the video in its entirety before proceeding.

www.solution-tree.com/Exploring_Different_Ways_That_Lines_Can_Intersect

In this video, students are asked to first predict which pairs of lines will be parallel and which will be intersecting. Notice that the task posed has multiple pairs of lines that will satisfy each conclusion. The teacher asks the students to justify why they think lines are intersecting or parallel from the equations and then compare those to graphs of pairs of the same lines. What conclusions are students able to make? Rate of change is connected to the equations and the graphs to make sense of the mathematics involved. Students conclude that when slopes are equal, the lines can be parallel or the same line, and when slopes are different, the lines intersect.

In this case, students are engaged in Mathematical Practice 7, "Look for and make use of structure." The equations of the lines provide a mathematical structure to make sense of how two lines are the same or different. Examining the slope and y-intercepts of each line provides a connection to the structure of the graphical representations, and conclusions are made based on this structure.

In the next video, you will explore high school students responding to the task provided in figure 1.1 (page 16). Watch as they interpret graphs that involve a race between father and son. You are encouraged to watch the video clip in its entirety before proceeding.

www.solution-tree.com/Interpreting_Graphs
_to_Create_Context

Traditionally, the focus of instruction on functions is limited to generating graphs from tables and equations. This video demonstrates that it is also important for students to interpret graphs accurately and create contexts that match graphs. Notice the questions the teacher asks as she circulates among groups. How do her questions support student thinking? She does not give answers early on, nor does she provide strategies. She asks questions that require the *students* to explain their thinking. The focus is on the students and how they are making sense of the problem. Asking questions like "What are you doing?" provides a way for the teacher to facilitate student thinking without taking the focus off of the students. When students ask questions or respond to the teacher's questions, she replies by asking additional questions, such as "Where are they when the race starts?" or "Does the father's pace change?" The impetus for these questions remains on facilitating student thinking.

A common misconception related to distance-time graphs arises in this discussion. This common misconception involves interpreting distance-time graphs as representing the situation literally. Students interpret the positive slope to mean speed is increasing instead of the distance. Because the graph is getting "higher," the person must be running faster. How does the teacher identify this misconception? What does she do the first time it is discovered? Initially, when asking a group to tell her what they are doing, one of the students says that the father and son meet at four seconds. Then the student determines that the son starts to go faster and the father slows down. In another group, discussion ensues about the same concept. One student says the father and son are accelerating, and another student disagrees, stating that the lines are straight, so their pace is constant. This same student, however, then goes on to say that after four seconds, the son starts to run faster.

What in the graph leads to these statements? At the start of the race, the father's graph is "above" the son's graph. At four seconds, the graphs of the father's and son's lines intersect. At that point in time, the son passes the father, so his distance from the start is greater and his graph is above the father's graph. This leads to misinterpreting the graph as the son is now running faster since his graph is above the father's graph. The teacher is aware of this misconception and uses a formative assessment process to see if it is present with the students. Since it appears in multiple groups, she uses whole-class discussion to help address this misconception. Through eliciting student responses in the whole-class discussion, the teacher is able to support students in making sense of what is actually happening in the graph. As the graph goes higher, it means the father and son have run farther, not faster. The students identify that the pace each was running had to be constant, even after the intersection point. This is important for students to understand. It would have been easy for the teacher to tell them this directly; however, by asking carefully crafted questions, students are able to make sense of it for themselves, and they are much less likely to make that error again. In this way, a common and anticipated error is used as a springboard for learning.

The teacher is fostering Mathematical Practice 2, "Reason abstractly and quantitatively," and Mathematical Practice 4, "Model with mathematics." By working on a task that is graphical in nature, students have to contextualize the problem by interpreting what was occurring at various points in the race. Students examine the graph to make sense of the context and then return to the graph to verify that their context is accurate. They also connect the graphical representation to a real-world scenario. This is especially important when examining distance-time graphs. Interpreting these graphs is an integral aspect of modeling the world around you.

TQE Process

At this point, it may be helpful to watch the second video (page 31) again and pay close attention to the tasks, questioning, and opportunities to collect evidence of student learning.

The TQE process can help you frame your observations. Teachers who have a deep understanding of the mathematics they teach:

- Select appropriate *tasks* to support identified learning goals
- Facilitate productive *questioning* during instruction to engage students in Mathematical Practices
- Collect and use student *evidence* in the formative assessment process during instruction

This lesson's *task* provides the opportunity for students to interpret linear graphs in a real-world situation. The task is rich enough to support discussions and reveal students' difficulties and misconceptions. For instance, the students might fail to distinguish between the slope and height of a graph. As a result, they don't realize that the speed of the runners remained constant. Sometimes students incorrectly infer that the runners will have the same speed at the intersection point. These misconceptions are uncovered because the task directs students to create the context. If the context was provided, the misconceptions could remain hidden. These misconceptions are revealed during class discussions, and the teacher uses the context of the problem to help students overcome their misconceptions.

The teacher uses *questioning* to gather information regarding how students are making sense of the problem. Perhaps you noticed the types of questions the teacher asks the small groups as she circulates among them. She asks questions like "What are you doing?" to encourage students to explain their thinking and their solution processes. What does the teacher do when one student answers for the group? She responds by asking additional questions to engage all members of the group in productive discussion and supports their interactions even after she moves to another group. Asking questions like "What does he mean?" provides ways for students to interpret and respond to each other's thinking. In one case, a student mentions that the son is accelerating faster than the father. The teacher asks the group what that means, which prompts another student to say, "That's what I wanted to know!" This provides an avenue for the students to engage in Mathematical Practice 3, "Construct viable arguments and critique the reasoning of others," by making sense of each other's thinking and productively discussing why ideas are correct or incorrect. The teacher's response of "Then you have a question for your group" also serves to promote productive talk. As she identifies which students display the common errors, the teacher knows how to address those errors with the class during whole-group discussion.

The teacher collects *evidence* of student learning through the formative assessment process. Through thoughtful questioning of students and their thinking, the teacher identifies several misconceptions related to the graph. These include misinterpreting the points on the graph and misinterpreting the slope as acceleration instead of speed. By asking pointed and purposeful questions, the teacher is able to gather formative assessment data that then drives the whole-class discussion. As the teacher brings the students back together for a discussion regarding this problem, she is able to use the data she collected in small groups to guide the discussion. She knows that some students misinterpreted what was occurring at the start of the race, so she is mindful to ask about it. Instead of providing a judgment of right or wrong, she prompts the students to justify their answers using the graph. She knows the interpretation of distance as speed is a misconception, so she makes sure to ask the whole class about the speed at various points in time on the graph, particularly after the intersection point. She is careful to elicit the misconception and to provide questions and discussion that help students determine what is correct.

The Response

It is well documented in the research literature (for example, Garcia & Cox, 2010; Janvier, 1981; Leinhardt, Zaslavsky, & Stein, 1990) that students often treat a graph as a picture of the situation described in the problem. For instance, in the task in figure 1.1 (page 16), students might confuse the shape of the distance-time graph with the speed of the runner or the type of path on which the runner is racing. As a result, they may assume that if the graph is constant, the runner is running on a flat horizontal surface. If the graph is increasing and decreasing, then the runner is going up and down. In this task, since the graphs are increasing from left to right, students might incorrectly infer that the runners run uphill during the race.

Additionally, students often have a misconception related to the meaning of the slope and the point of intersection of the graphs. In the task in figure 1.1, between A and B, the graph of line F is "higher" than the graph of line S. Does this mean that the speed of the father is greater than the speed of the son? As occurred in the video, students who fail to make the distinction between the slope and height of a graph tend to think that the father is running faster than his son between A and B and that they have the same speed at the intersection point. The intersection of the lines typically indicates where and when the runners meet. However, this is not always true for distance versus time graphs. If the runners had been on opposite sides of the track, for example, and had run to the same finish line, then the intersection would represent when the runners are at an equal distance from the finish line. Some students incorrectly infer from the graphs that the runners pass each other or will have the same speed at the intersection point. By presenting different graphs of lines (see figure 1.1) and allowing students to write stories, students make connections between the shape of the graphs and the context to overcome their difficulties. Students should be given opportunities to create contexts for a variety of graphs and to create graphs to model contexts so that the relationships between contexts and shapes of graphs become explicit.

Reflections

1. What do you feel are the key points of this chapter?
2. What challenges might you face when implementing the key ideas from this chapter? How will you overcome them?
3. What are the important features for developing an understanding of equations and functions, and how will you ensure your instruction embeds the support needed for these features?
4. Select a recent lesson you have taught or observed focused on equations and functions. Relate this lesson to the TQE process.
5. What changes will you make to your planning and instruction based on what you read and considered from this chapter?

CHAPTER 2

Structure of Equations

The initial focus of this chapter is solving systems of two linear equations. The discussion will then focus on exploring different forms of linear and quadratic functions and linking expressions and equations to their visual representations. This exploration sets the stage for later work with other types of functions.

The Challenge

The initial task in this chapter provides an application of linear functions (see figure 2.1). This application is an extension from the context presented in chapter 1 (see figure 1.1, page 16).

DO NOW

The figure below illustrates a distance-time graph for a father (F) and son (S) during a 100-meter race. Use the information provided to:

1. Write an equation to represent each line
2. Determine who is running faster
3. Identify the distance from the finish line where the father and son meet

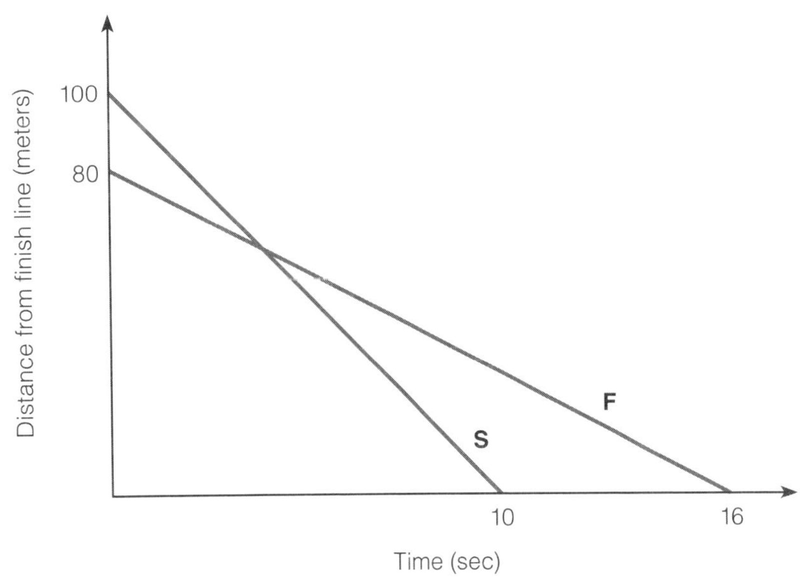

Figure 2.1: A race between a father and son.

In this task, the two linear graphs representing distance over time intersect at a single point, indicating that the system of equations has a unique solution. In high school, there has traditionally been an emphasis on two methods for finding the solution of a system of linear equations—substitution and elimination. These methods don't always address the meaning of the solution to a system of equations.

What does the solution mean within a given context? This chapter explores various methods to make sense of solving systems of equations to illustrate the mathematical reasoning behind why different methods work and how they are interconnected graphically and algebraically. What does the solution mean graphically? What does a solution to a system of equations mean algebraically? You may wish to consider these questions before reading the discussion on graphical and algebraic solutions of the initial task in the following paragraphs.

What does a point of intersection mean for a system of linear equations? What are the possibilities for the number and type of solutions for two linear equations like those in figure 2.1 (page XX)? Recall that the set of solutions for a linear equation in two variables can be represented by the set of unique points on the Cartesian coordinate plane that satisfy the equation. The collection of all possible points that satisfy a linear equation determines a line. From the graphs of the two lines in the task in figure 2.1, you see the father is given a 20-meter head start. This is shown by the y-intercept of the line showing that the father is 80 meters from the finish line, while the son is 100 meters from the finish line. The son runs at a faster rate and passes his father at some point in the race. How can you determine when and where the father and son meet? While you can estimate the point of intersection from the graph, you may want to determine that point with greater precision. How do the equations of the lines help you think about this situation? Consider how the equations of the lines illustrate distance traveled over time. Since the x- and y-intercepts of each line are given, you may choose to determine the slopes of each line using that information. Let m_F be the slope of the line representing the father's distance over time, and m_S stand for the slope of the line representing the son's path, in meters per second:

$$m_F = \frac{80 - 0}{0 - 16} = -5$$

$$m_s = \frac{100 - 0}{0 - 10} = -10$$

What did you notice about the rates of change for the father and son? Why are the slopes negative? What implications do these numerical slope values have in the context of the race? The values of the slopes are both negative, showing that the distance between the runner and the finish line is decreasing. Since slope is a measure of the rate of change, the context of the problem allows you to state that the slope measures how fast each runner is going. Who is moving faster? This is an interesting question, as the numerical value of the rate of change seems to suggest that the father is moving faster since $-5 > -10$. However, the negative sign is actually indicating direction, not the speed of the movement. As a measure of distance per unit of time, you need to consider the absolute value of the slopes to determine that the son is moving faster toward the finish line at 10 meters per second compared to the father's 5 meters per second. This is an important consideration of how the context of the problem influences how you interpret the mathematics of the situation.

What form of linear equation did you write for each line? You could have chosen the point-slope, slope-intercept, or standard form as shared here respectively:

Equations for father's line: $y - 80 = -5(x - 0)$, $y = -5x + 80$, or $5x + y = 80$

Equations for son's line: $y - 100 = -10(x - 0)$, $y = -10x + 100$, or $10x + y = 100$

Each of these equations expresses different ways to show the relationship in the task. You might become aware that you tend to select one form of the line over another based on habit. You should remain aware of these choices so that you are intentional about allowing students to make their own choices and then ask them to defend their selections. The different forms will be discussed later in this chapter.

Continuing the examination of the point where the father and son meet, consider how substitution can be used to determine the point of intersection. Since the *x*- and the *y*-values for both lines are the same at this point, you can substitute the value of *y* from one equation into the other. The slope-intercept form of the line is the most direct form of the line to use for this strategy since both equations are already solved for *y*. The expression $-5x + 80$ from the equation for the father can be set equal to $-10x + 100$ from the equation for the son so that you have a new equation in one variable, $-5x + 80 = -10x + 100$. Each side of this new equation is represented as a graph of *y* in terms of *x* in figure 2.1 (page 35). The next steps in solving for *x* could be:

$$5x + 80 = 100$$

$$5x = 20$$

It is important to emphasize that each step in the solution process results in a new equation as the same quantity is added to both sides of the previous equation, thus maintaining the equality. A new set of two lines can be graphed to correspond to each step in the solution process. These new graphs no longer represent the context of the problem in that the father and son's races are not evident in the graphed lines but the graphs do serve as a visual representation of the solution. This use of abstract thinking to make sense of quantities supports engagement in Mathematical Practice 2, "Reason abstractly and quantitatively."

Consider the graphs in figure 2.2 (page 38). What do you notice about the point of intersection in each graph? You probably noticed that each of these graphs of systems of equations has the same *x*-value for the point of intersection, even though the *y*-value is different for each point.

What if you simplify the equation $5x = 20$ further? The final equation of $x = 4$ results in the graph of a vertical line. What are the conceptual implications of this vertical line to the system of equations? Notice that this vertical line tells you that the point that satisfies both linear equations lies on this vertical line $x = 4$. Although there are infinitely many points on this vertical line, all points have the same *x* value. Thus, you can conclude that the *x*-coordinate of the point that solves both equations is equal to 4 and that the father and son will meet 4 seconds after the start of the race. Making sense of the graphical solution in terms of the context of the problem represents the quantitative reasoning of Mathematical Practice 2.

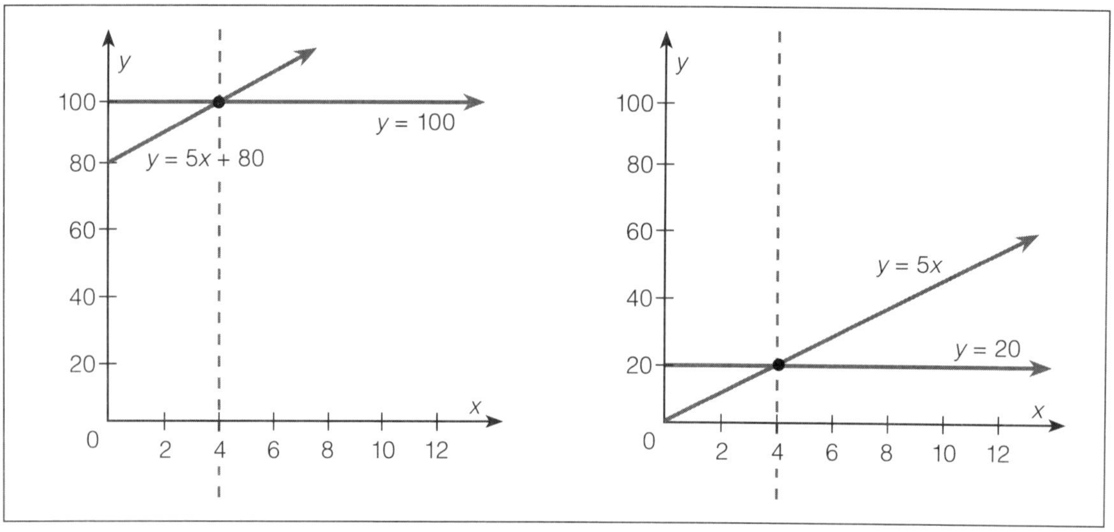

Figure 2.2: Graphs of the solution process to a system of linear equations with substitution.

Conceptually, when each of the equations is solved for y prior to the substitution, you are representing the father's and son's distances in terms of their times. When the substitution is made, the assumption is that this distance is the same for each runner. What would be the meaning of the context if both equations were solved for x? If you solve for x, you are representing how long the father or son has been running in terms of his distance from the finish line. Now when you substitute, you are assuming that the amount of elapsed time is the same and you are ascertaining the distance where the father and son meet. Substituting for x will require algebraic manipulation regardless of the initial form of the line you choose. Algebraically, if you solve the equation for the father's line for x and then substitute the expression $16 - y/5$ for x in the second equation, you will have:

$$y = -10(16 - \frac{y}{5}) + 100$$

$$y = -160 + 2y + 100$$

$$y = 2y - 60$$

$$-y = -60$$

$$y = 60$$

The graph of the equation $y = 60$ is a horizontal line. Since all points on a horizontal line have the same y-coordinate, you know that the y-coordinate that satisfies both equations is 60 and that in the context of this problem, the father and son will meet 60 meters from the finish line. Again, if you consider the expressions on both sides of the equation $y = 2y - 60$ as separate functions and use their graphs to solve the system of linear equations, you will see that the y-coordinate of their intersection point remains the same as for $y = 60$ (see figure 2.3).

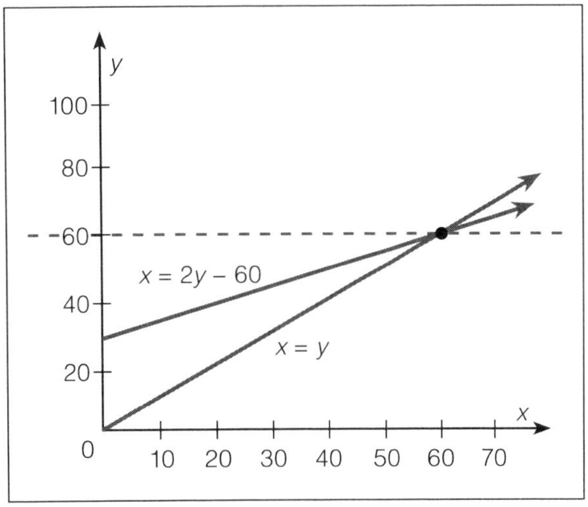

Figure 2.3: Solving a system of linear equations with substitution.

What conclusions can you draw regarding the horizontal line drawn at $y = 60$? Through observation, you may notice that when a new equation in one variable is created by the substitution method, the resulting graph is either a vertical line or a horizontal line. Solving both systems of equations provides you with the coordinates of the point of intersection.

Next consider a technique that is often used when the equations are not initially solved for y or x—the method of elimination. The standard forms of the equations for the father and the son will be used to illustrate this method.

$$5x + y = 80$$

$$10x + y = 100$$

In order to eliminate one of the variables in this example, you can rewrite one of the equations. For instance, multiplying the first equation by –1 will eliminate the variable y when the equations are added. In order to add these equations, you add the x terms to the x terms, the y terms to the y terms, and the constants to the constants. The resulting equation will be $5x = 20$, which simplifies to $x = 4$.

$$\begin{aligned} -5x - y &= -80 \\ \underline{10x + y} &\underline{= 100} \\ 5x \quad &= 20 \\ x \quad &= 4 \end{aligned}$$

At this point, you could use this x-value to substitute into one of the original equations and solve for y.

On the other hand, you could choose to multiply the first equation by –2 to eliminate the variable x when the equations are added.

$$\begin{array}{r} -10x - 2y = -160 \\ \underline{10x + y = 100} \\ -y = -60 \\ y = 60 \end{array}$$

How does the method of elimination connect to the graphical solution? In both combinations, one line was not changed, $10x + y = 100$. In both instances, the other line was manipulated to be either $-5x - y = -80$ or $-10x - 2y = -160$. Regardless, the graph of this line will overlay the original graph for $5x + y = 80$, the father's line.

Why does the elimination method result in a solution to the system of equations? The algebraic identity "if $a = b$ and $c = d$, then $a + c = b + d$," or the addition property of equality, is often used as the mathematical justification for why the elimination method works. In other words, since both sides of each equation are equal, you can add sides from different equations to those sides. However, it is not clear to many students why this method works graphically. The manipulated forms of the lines graph the same lines, and therefore, the values of the x and y variables represent the coordinates of the point of intersection for the lines $x = 4$ and $y = 60$ (see figure 2.4).

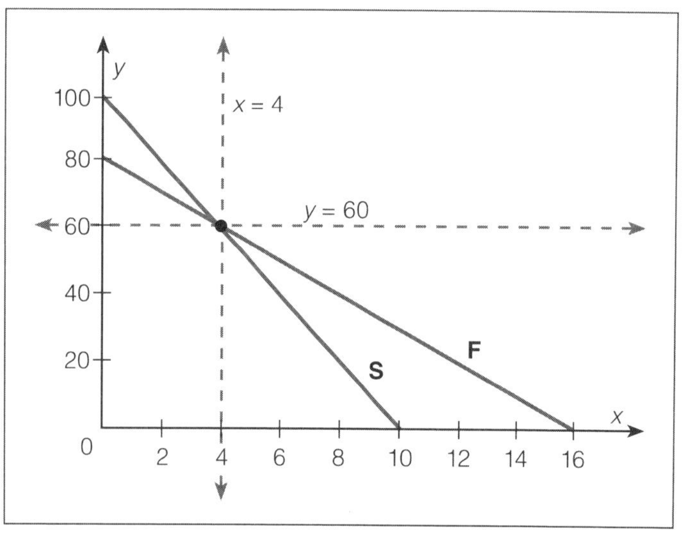

Figure 2.4: Solving system of linear equations with elimination.

The ability to understand the impact of and connection between graphic and algebraic changes is important for students to develop. A flexible understanding of equality and the various ways in which quantities can be altered to produce equivalent expressions is at the heart of making sense of mathematics in algebra and beyond.

The Progression

A conceptual understanding of linear functions is important in mathematics because many natural phenomena can be modeled by linear equations. In the middle grades, students learn to formulate linear equations in order to solve word problems. In grade 8, systems of linear equations are introduced, and in high school, the focus shifts to solving real-world problems represented by two linear equations in two variables. In high school, students also write equations in different forms for linear and quadratic functions and explore advantages of each form in different situations. The following provides an outline of a progression for the development of a conceptual understanding of linear and quadratic functions and systems of linear equations in high school.

- Construct linear and quadratic equations in two variables.
- Analyze linear and quadratic functions using multiple representations.
- Use linear and quadratic functions to model real-world situations.
- Explore different forms of linear and quadratic functions.
- Solve systems of equations using multiple strategies.

Grade 8

Throughout the middle grades, students connect proportional relationships, lines, and translations to linear functions. They know that a function represents a rule that applies to all points on its graph, and thus, the graph of a function is determined by a collection of ordered pairs that satisfy the equation. For instance, given the problem "A man is traveling at the speed of 5 miles per hour. How many miles will he travel in 1, 2, 3, or x hours?" students recognize that the speed is constant, and there is a proportional relationship between distance and time. The relationship between distance and time can be expressed by a linear equation, $y = 5x$. In grade 8, students know that the linear equation $y = 5x$ states a rule that the y-coordinate (or distance) of a point on the line is five times the x-coordinate (or time). As they construct and solve linear equations, students describe and explain what a solution represents in multiple representations. As they explore the relationship between distance and time, they create a table of values and plot points on the Cartesian coordinate plane. Students learn that every point on the graph of the line represents a solution for this equation, and a visual representation of the solution for this linear equation creates a straight line. When presented with a system of two linear equations, they understand that a possible solution is the point of intersection of their graphs because the point of intersection satisfies both equations simultaneously. In grade 8, students also extend their exploration of functions to define, evaluate, and compare linear and nonlinear functions using different representations.

High School

In high school, students extend the concepts related to linear functions to explore different forms of linear functions beyond the slope-intercept form of $y = mx + b$. As students gain experience with algebraic equations and their solutions in different representations, they come to understand the advantages of writing equations in various forms. Seeing structure in equations and exploring the benefits of each structure is an important component of working with functions in different forms as students understand and express functional relationships in multiple representations.

Throughout high school, students extend their interaction with linear equations to include representing and solving systems of equations graphically. Following explorations from grade 8 when students analyzed and solved linear equations and solved systems of linear equations, high school students make sense of the solution of systems of linear equations and connect multiple representations of systems of equations to one another. In addition, they prove that the method of substitution or elimination produces a system with the same solutions.

Students explore quadratic functions in the standard, factored, and vertex forms. They derive the vertex form and the quadratic formula by completing the square so that they can apply these forms to represent, understand, and solve real-world and mathematical problems. Students will further explore and apply quadratic functions and other functions in later high school work.

The Mathematics

The ability to represent both real-world and mathematical situations in many different ways is important to solve problems. Students develop flexibility in the ways they mathematize situations by exploring multiple representations of functions and by justifying the procedures they use. Learning to create and use these representations is important in solving problems efficiently. High school students must be able to model contexts through visualization, explore different forms of linear functions, and explore different forms of quadratic equations.

Modeling Contexts Through Visualization

Students learn to model contexts with mathematics in elementary school and the middle grades. This often occurs in ways that are different than the ways that many teachers learned to make sense of these situations. Consider the middle school bridge task provided in figure 2.5. How would you solve it?

DO NOW

How many beams are in a bridge of length 4? How about a bridge of length 5? 10? How many beams are in a bridge of length *n*? How do you know?

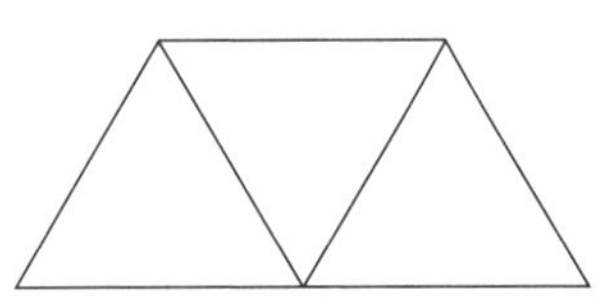

Bridge length: 2 beams
Total number of beams in bridge: 7 beams

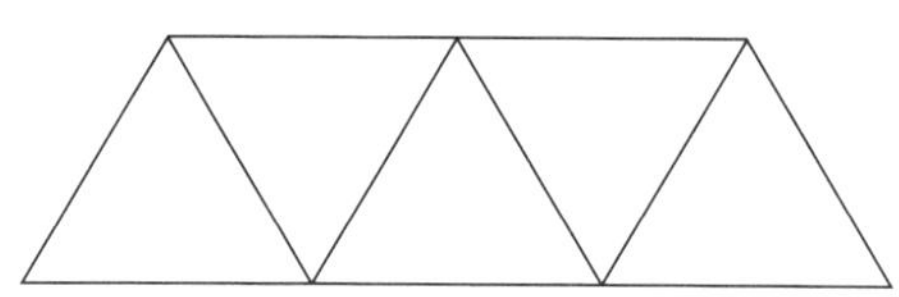

Bridge length: 3 beams
Total number of beams in bridge: 11 beams

Figure 2.5: Bridge task.

Often teachers reason in ways that mask some aspects of a mathematical relationship. For example, you might make a table to solve this problem similar to the one provided in table 2.1. How could you use this table to find the number of beams in a bridge of length n?

Table 2.1: Table of Values for Bridge Task

Bridge Length	Total Number of Beams
2	7
3	11
4	15
5	19

The generalization, or the expression to find the bridge of any length, could be determined recursively by noticing the change from one term to the next. The number of beams increases by a constant rate of four beams from one bridge to the next. However, if you multiply the bridge length by 4 for each bridge, the number of beams is one too many for each bridge. This leads to the generalization of $4n - 1$ where n is the length of the bridge and the expression $4n - 1$ is used to determine the total number of beams.

This method for determining the generalization for the number of beams in a bridge of any length is fine; however, one limitation is that the total number of beams is tied exclusively to the use of the numeric table representation. The meaning of the context and the benefits of making sense of the context are masked in the development of the expression. It may be difficult to connect each part of the expression back to the context. Instead of using a table to determine the generalization, what if you reasoned visually? This task may be challenging if you did not use the images of the bridge to determine a generalization. How *could* the visual of the bridge lead to a generalization? Consider the visual provided in figure 2.6.

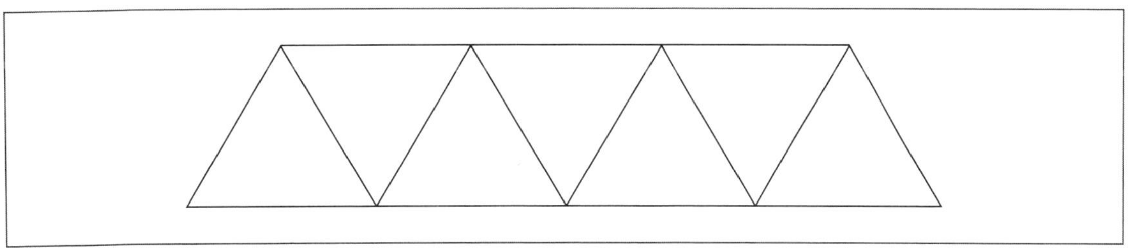

Figure 2.6: Visual for bridge task.

How can you generalize the pattern of the number of beams? How do you connect the visual with the generalization? Examine each of the generalizations provided in figure 2.7.

$3 + 4(n - 1)$

$n + 2n + (n - 1)$

$3n + (n - 1)$

Figure 2.7: Visualizing expressions for the bridge task.

Each generalization is equivalent to $4n - 1$, but it was not determined using a table. Rather, each expression connects to a particular way of *visualizing* the bridge. How does the image of the bridge connect to each expression? In making sense of the expression $3 + 4(n - 1)$, what does the 3 represent in the visual? What about the 4? Why does the expression use $(n - 1)$ rather than n?

Where does each term in the expression $3 + 4(n - 1)$ connect to the visual in figure 2.6? The 3 can be seen in the first triangle. The 4 represents the number of beams added on to the bridge each time the length of the bridge increases by 1. The $(n - 1)$ can be more difficult to see. It represents the length of the bridge minus the first triangle. This is because the first triangle uses three beams, and each time the bridge length increases by 1 *after* the first triangle, 4 additional beams are added (see figure 2.8).

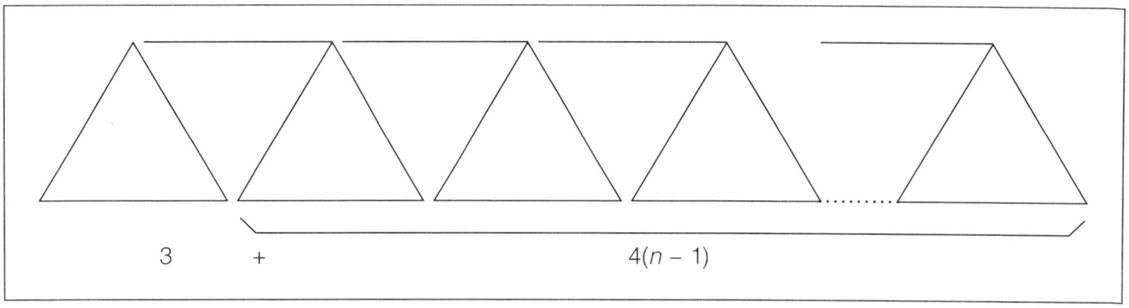

Figure 2.8: Visual for 3 + 4(*n* – 1).

Now that you have explored how an expression can connect to the visual image, revisit the second expression in figure 2.7 (page 43). How does $n + 2n + (n - 1)$ connect to the visual in figure 2.6 (page 43)? To what part of the picture does the n connect? What about the $2n$ and the $(n - 1)$?

The n refers to the number of beams in the length of the bridge, or in the visual, the number of beams along the base of the bridge. The $2n$ refers to the beams that connect to the base of the bridge. The number of beams across the top of the bridge is 1 fewer than the number of beams along the base; therefore, that number can be modeled by the term $n - 1$.

How about the expression $3n + (n - 1)$? This expression is probably easier to *see* in the visual in figure 2.6 because of your work with the previous two expressions. The more you explore using a variety of representations to mathematize situations, the better you get at their use. The same is true for your students. In this expression, the 3 refers to the number of beams in one triangle; it is multiplied by n to indicate that the triangles are those that have a side that makes up part of the base of the bridge (see figure 2.9).

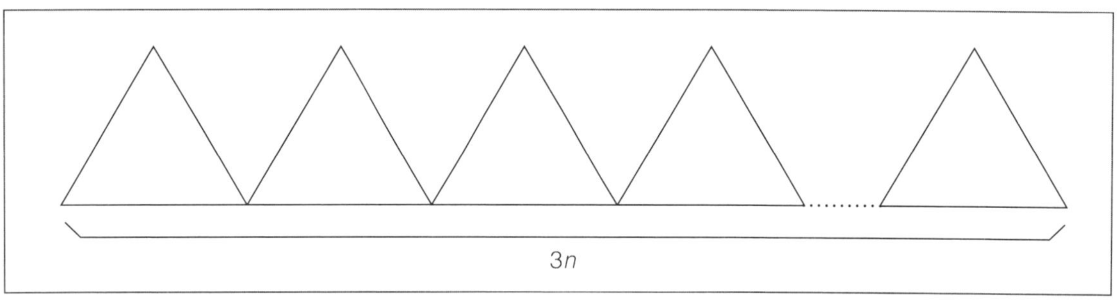

Figure 2.9: Visual for 3*n*.

The last term in the expression, $(n - 1)$, describes the number of beams that connect the triangles at the top of the bridge. As in the previous expression, this makes sense because the number of beams at the top of the bridge is 1 fewer than the number at the base.

The context of the problem is valuable in making sense of the expression. When students model with mathematics, it is important for them to refer back to the situation to make sure that their models are appropriate. However, sometimes the situation can lead directly to the generalization. In the bridge task, the expression was implicit in the visual model and it was not necessary to create a table to determine the generalization. When you are too quick to go to a procedure, like the process of making a table, you might miss an opportunity to *see* the mathematics in a more efficient way. Students may not always see a problem in the same way as either you or other students. Be sure to provide opportunities for them to share their thinking and make sense of the thinking of others. In doing so, you are supporting students to engage in Mathematical Practice 3, "Construct viable arguments and critique the reasoning of others." Encouraging students to make sense of problems on their own should also take place when exploring different forms of linear functions.

Exploring Different Forms of Linear Functions

In high school, different forms of equations are used to represent linear functions, such as $y = mx + b$, $y - y_1 = m(x - x_1)$, and $Ax + By = C$, but these different forms are not of optimal benefit unless the use of these forms also includes an explanation of which form is most appropriate in given situations. Examine the form $y = mx + b$. What critical information does this form of the equation of the line provide? What are the presumptions it makes regarding the position of the line? What type of values can the slope of the line take on in this form of the equation of the line? Notice that the linear equation of the form $y = mx + b$ relates x as the independent variable, y as the dependent variable, and m and b as constants. Hence the value of the y-intercept is b and the value of the m determines the slope of the line. The greater the absolute value of the slope, the steeper the line. The equation of the form $y = mx + b$ is particularly useful as this form can represent all nonvertical lines that may or may not pass through the origin, and given the slope and the y-intercept, the equation of a line can be easily constructed.

How does the slope formula connect to the point-slope form of the equation of a line? Consider the slope formula $m = \frac{y_2 - y_1}{x_2 - x_1}$. What would happen if instead of two fixed points (x_1, y_1) and (x_2, y_2), you examined the line created by a point (x_1, y_1) and a generic point (x, y) anywhere on the line? If you are given the slope and one point on the line (x_1, y_1), then the formula for the slope of the line would be $m = \frac{y - y_1}{x - x_1}$. When you rearrange the new equation, you derive the point-slope form of linear functions: $y - y_1 = m(x - x_1)$.

What is the relationship between this form and the previous $y = mx + b$ form? What would happen if, instead of a fixed point (x_1, y_1), the given point were to be on the y-axis? As long as the line is not vertical, this change in point would not result in a loss of generality and you could call this point on the y-axis $(0, b)$. What would happen if you substitute the point $(0, b)$ in for (x_1, y_1)? The equation transitions from $y - y_1 = m(x - x_1)$ to $y - b = mx$, which can be written as $y = mx + b$. Establishing these interconnections between various forms of the equations of the line can be quite empowering to demonstrate the coherence and the logical structure of mathematics, an example of Mathematical Practice 7, "Look for and make use of structure."

What happens when there is a slope that cannot be represented? What does this mean mathematically? To explore this question, consider the equation of the line passing through the points (3, 1) and (3, 6) using one of the forms of the equations of a line previously discussed—slope-intercept form $y = mx + b$ or the point-slope form $y - y_1 = m(x - x_1)$. What numerical value did you get for the slope in order to substitute it for the parameter m in these formulas? Using the two given points, you would get m equals $5/0$, but this is an indeterminate value, so the slope is undefined. How do you deal with lines that have undefined slopes? Because vertical lines have undefined slopes, they cannot be represented with either the slope-intercept or the point-slope form of a line. Lines with undefined slopes can be represented using the standard form of a line, $Ax + By = C$, as the slope is not explicitly stated in this form. In this example, you substitute in the two points (3, 1) and (3, 6) into the equation $Ax + By = C$ to get $A(3) + B(1) = C$ and $A(3) + B(6) = C$, or $3A + 1B = C$ and $3A + 6B = C$, respectively. Since the C values are the same, then $3A + B = 3A + 6B$, or $0 = 5B$, so $B = 0$. As the B term is equal to zero, $Ax + By = C$ will take on the form $Ax = C$. For the equation of the line passing through the points (3, 1) and (3, 6), the equation will be $x = 3$.

One advantage of using the standard form of a line is that any line in the Cartesian coordinate plane—including vertical lines—can be described by this form. Through exploration, students come to see the benefits of certain forms of equations of a line over others. At times, the format is determined by the givens in the problem, and at other times, the format is driven by the goals of the problem. The same is true for quadratic equations.

Exploring Different Forms of Quadratic Functions

By definition, a quadratic function is a function that can be described as $f(x) = ax^2 + bx + c$, where a, b, and c are real numbers and a is not equal to zero. This form of the quadratic function is commonly referred to as the standard form. It is helpful for students to explore the constraint on the value of a being nonzero to determine the form of a quadratic function if the value of a is zero. As you may have realized, the function will be a linear function if the value of the parameter a in the standard form of a quadratic function is zero.

Students explore quadratic functions in several forms. Why is this useful? Consider the task provided in figure 2.10 explore additional forms of the quadratic function.

DO NOW

The spray from a water fountain is modeled by the quadratic functions $f(x) = -(x - 5)^2 + 25$ and $f(x) = -x^2 + 10x$.

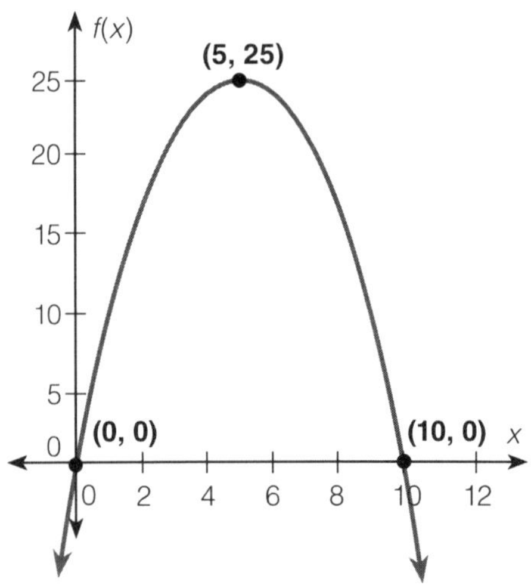

Which form of the quadratic function would you use to determine the following information if the graph was not provided? Why would you choose that form?

1. The maximum height reached by the spray.
2. The distance between the starting point and the ending point of the spray.

Figure 2.10: Model using different forms of a quadratic function task.

While both functions model the path of the water in this fountain, which mathematical form is more suitable to answer the questions posed in this task? Which function, $f(x) = -(x - 5)^2 + 25$ or $f(x) = -x^2 + 10x$, allows you to more efficiently identify the vertex based on its structure? From the graph, you see that the maximum value occurs at (5, 25). In examining the two different representations of the quadratic function given in the task, you see that 5 and 25 both appear in the first equation. In this task, the value (5, 25) can be identified as the vertex (h, k) yielding the highest point on this curve. This equation is the vertex form of the quadratic equation, represented by $y = a(x - h)^2 + k$. The values for h and k identify the vertex, but why? It is helpful to provide students opportunities to see that this form of the quadratic function *always* identifies the extreme value—maximum or minimum—of a quadratic function. This can be accomplished by engaging students in Mathematical Practice 8, "Look for and express regularity in repeated reasoning," by providing several graphs of functions along with the vertex form of the equation for students to examine. Once students are able to conjecture that the vertex is identified within the vertex form by (h, k), they can be supported to engage in Mathematical Practice 7, "Look for and make use of structure," to determine why this is the case. If given the opportunity, they can see that $(x - h)^2$ cannot be negative and the minimum value for $(x - h)^2$ occurs when $x - h = 0$. That minimum value describes the turning point of the graph, and at that minimum value, k describes the value for y because when $x - h = 0$, the function becomes $y = k$. An interesting question to ask students is how this minimum value can also describe the maximum point of the function, as with the case in the task in figure 2.10. Students come to see the importance of the value of the parameter a in the vertex form to determine whether the vertex describes a maximum or minimum point of the function.

Which form of the quadratic function provides information about the starting and ending points of the spray? How can you determine those values from the given information? You see from the graph that the water fountain spray has a height of 0 at the x-coordinate values 0 and 10. You probably notice that there is a 10 in the second equation and wonder how this might be valuable to determine the starting and ending points. The second equation is actually the standard form of the quadratic equation, and a quick analysis enables you to identify $y = -x^2 + 10x = -x(x - 10)$. This provides you the factored form of the quadratic equation. From the factored form of the equation, it is more apparent that the values of $x = 0$ and $x = 10$ represent the starting and ending points, which allows you to identify the distance between these points as 10 units. In summary, the vertex form provided the needed structure to identify the maximum height, and the standard form of the quadratic function, $y = -x^2 + 10x$, allowed you to employ various factoring techniques to find the intercepts.

Quadratic functions provided in standard form can be changed into the factored form $(x + m)(x + n)$ in order to determine its zeroes (where the graphs of quadratic functions cross the x-axis). How do you transform the standard form of the quadratic function to the vertex form to determine the maximum or minimum value of the quadratic function? A commonly used technique is known as *completing the square*. This can be accomplished geometrically or algebraically.

The geometric exploration makes use of algebra tiles. Quadratic expressions can be represented as the "area" of a rectangle using algebra tiles with the factors of the quadratic expression as the dimensions of the rectangle. Area is used loosely here because the negative values of terms are also described as part of the area even though space cannot be negative. Each of the tiles is represented by its dimensions, and shading can indicate whether the term is positive or negative. The tiles representing the negative terms are designated in a different color or shading and are subtracted from the area rather than added. The standard form of the quadratic function provided in the task in figure 2.10 (page 46) has a coefficient of –1 for the x^2 term. When completing the square, it is common to factor out the coefficient of this term if it is something other than 1. Returning to the equation representing the water fountain spray $y = -x^2 + 10x$, factoring out –1 leads to $-x^2 + 10x = -(x^2 - 10x)$. When representing $(x^2 - 10x)$ with algebra tiles, x^2 is positive and the ten x tiles are represented as negative values (see a in figure 2.11).

When displaying a quadratic function using an area representation, it is helpful to begin with the squared term and then the x term. An algebra tile in the shape of a square with side lengths equal to x represents x^2, and a rectangle with dimensions x by –1 represents each $-x$. The goal of representing the function as a square determines how these terms are arranged as tiles (see b in figure 2.11). Remember that a rectangle with negative dimensions is acceptable with algebra tiles.

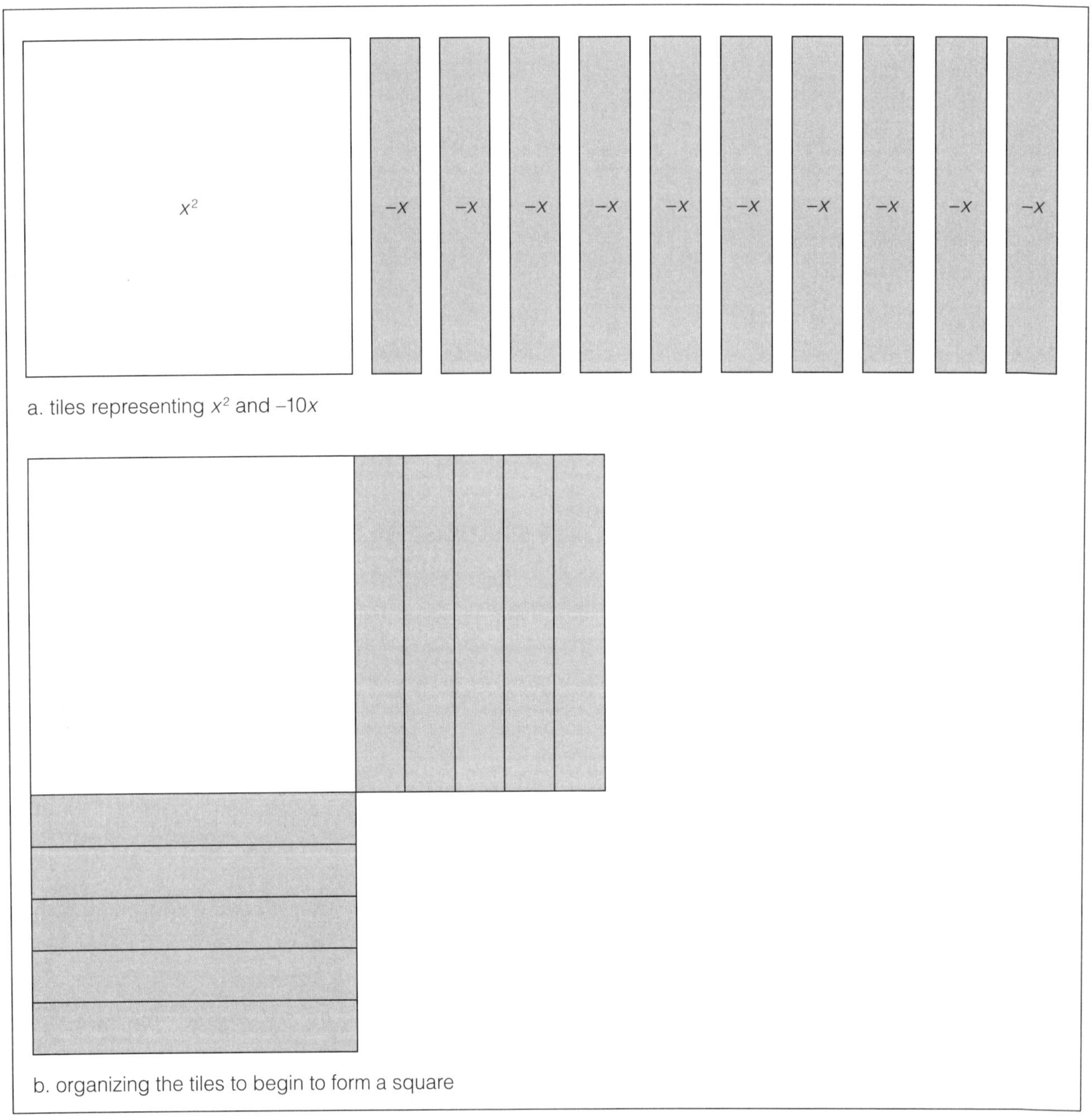

Figure 2.11: Representing $-x^2 + 10x$ geometrically.

The goal is to *complete the square*. What is missing? Twenty-five 1-by-1 squares would fill the gap to complete the square (see figure 2.12). Note that these 1-by-1 squares are positive, as they represent the –1 by –1 dimension from the $-x$ tiles. This represents what is added to the expression $x^2 - 10x$ to complete the square.

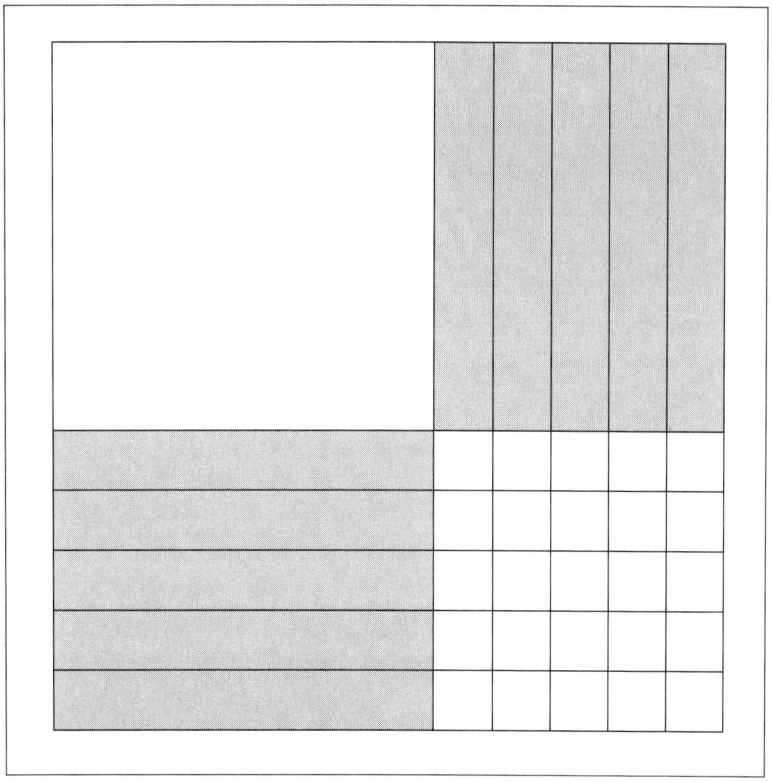

Figure 2.12: Completing the square geometrically.

In order to illustrate the area of the newly created square, you would have $(x - 5)^2$. However, this binomial includes the 25 ones that were added to complete the square. In order to maintain equality with $x^2 - 10x$, those ones need to be subtracted so that in essence, 25 is both added to and subtracted from the expression, leaving the overall expression unchanged. However, since the –1 was factored out of the expression, you now have $-x^2 + 10x = -[(x - 5)^2 - 25] = -(x - 5)^2 + 25$.

$$x^2 - 10x = (x - 5)^2 - 25$$

$$-(x^2 - 10x) = -[(x - 5)^2 - 25]$$

$$-(x^2 - 10x) = -(x - 5)^2 + 25$$

This manipulation results in the vertex form for the function expressed as $f(x) = -(x - 5)^2 + 25$. Notice how this solution process relates to the bridges problem in figure 2.5 (page 42) in that you are able to *visualize* the terms in the expression but this time with algebra tiles.

Why, in the beginning of the process of completing the square with algebra tiles, was the –1 factored from the x^2 term? Why wasn't $-x^2 + 10x$ used to complete the square? This is an interesting exploration. Consider the square determined by $-x^2$. Remember that with algebra tiles, negative dimensions can be modeled using shaded tiles. Because of this, it is possible to have a square that has both positive and negative dimensions to create negative area. In order to model $-x^2$ with algebra tiles, the square would need to have dimensions of x and $-x$ (see figure 2.13).

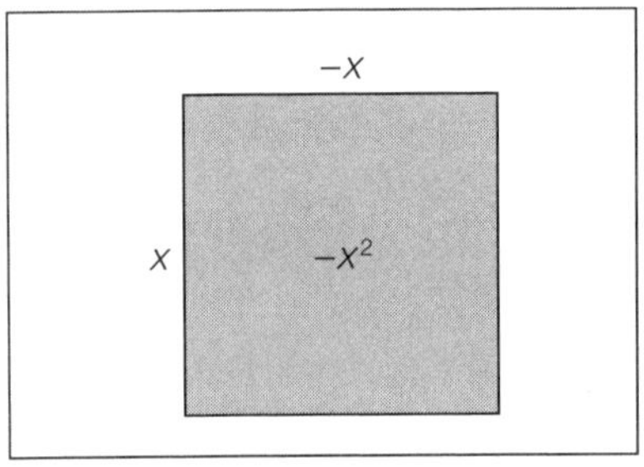

Figure 2.13: Representing $-x^2$ using algebra tile.

This is probably a bit concerning, this idea of having a square without side lengths that are equal in measure. However, this is required to model negative terms with an area model. It is an "open" area model since this negative space cannot really be modeled geometrically. However, algebra tiles serve an important purpose in helping to make sense of multiplying binomials and factoring trinomials, so this practice of using them to model negative space is accepted by many.

Moving forward with this idea, how would the ten x tiles be situated with the $-x^2$ term? The x terms have dimensions of x and 1. These dimensions need to be both positive or both negative so that the value of the x tile is positive. Figure 2.14 provides the dimensions necessary to set up the arrangement to complete the square.

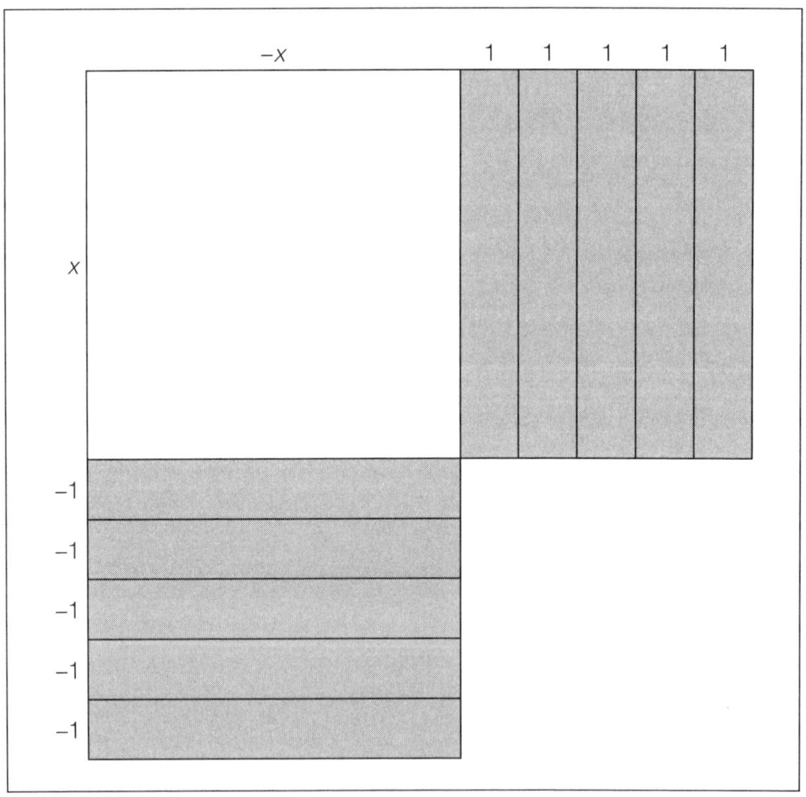

Figure 2.14: Representing $-x^2 + 10x$ with algebra tiles.

The x tiles with the dimension of –1 are necessary because the side they share with the $-x^2$ tile is represented by $-x$. The larger square would need to be completed with a region that is described as –5 by 5 so that the region completing the square would be –25. What would be the dimensions of the larger square? They would be $(x - 5)$ and $(-x + 5)$ or $-(x - 5)$ so that the "square" would be described as $-(x - 5)^2$. Because the –25 was added to the area, 25 would need to be added to the expression as well so that the quadratic function is rewritten as $f(x) = -(x - 5)^2 + 25$, the same as when the –1 was factored out prior to starting the process of completing the square in figure 2.11 (page 48).

Explorations such as these are valuable for students so that they don't just take rules at face value. Some rules—such as using a coefficient of one for the x^2 term—are employed because they make modeling easier; others—such as splitting the x terms equally to both sides of the square—are used because they make modeling possible. Students who engage in Mathematical Practice 3, "Construct viable arguments and critique the reasoning of others," make sense of the difference.

How does the algebraic process of determining the solution relate to the visual process? With the algebraic process of completing the square, there is an assumption that the coefficient of x^2 is 1, so that a square will truly have sides that are each the same measure. The rule that dimensions of squares need to be positive and equal in measure is not suspended in the algebraic representation. Once that assumption is met by factoring out the coefficient if necessary, you take the square of half of the coefficient of x and add and subtract this number to the original expression. How is half of the coefficient of x related to the geometric representation provided in figure 2.12 (page 49)? Notice that half of the x terms make up part of one side of the square. By sharing the x terms fairly on each of two adjacent sides of the square, the sides are assured to be the same length. The area of the missing square in b of figure 2.11 is determined by squaring the half of the coefficient of the x term that makes up one side of the incomplete square. A shortcut to determine the area of the missing square is to take half the coefficient of the x term and square it. This is both added to and subtracted from the original expression so that the value of the original expression is not changed.

As an example, let's consider the same quadratic expression $-x^2 + 10x$. Since the coefficient of x^2 is –1, the –1 can be factored out so you have $-(x^2 - 10x)$; you add and subtract the square of half of the coefficient of x, which is $(^{-10}\!/\!_2)^2 = (-5)^2$ in this example.

$$-[x^2 - 10x + (-5)^2 - (-5)^2]$$

$$= -[x^2 - 10x + (-5)^2 - 25]$$

$$= -[(x - 5)^2 - 25]$$

$$= -(x - 5)^2 + 25$$

So far you have determined the zeroes of the function by factoring and made sense of how to determine the coordinates of the vertex by completing the square, both in the context of a real-world situation. The method of completing the square can also be used to obtain a general formula for solving any quadratic equation. You start with the standard form of quadratic functions, $f(x) = ax^2 + bx + c$. Then set the expression describing the function equal to zero and solve for x.

$$ax^2 + bx + c = 0$$

To align to the previous process, you can subtract c from both sides to maintain quality and divide all terms by a:

$$x^2 + \tfrac{b}{a}x = -\tfrac{c}{a}$$

Now the coefficient of x^2 is 1. To complete the square, you must add the square of ½ the coefficient of x to both sides of the equation and complete some algebraic manipulation:

$$x^2 + \tfrac{b}{a}x + (\tfrac{b}{2a})^2 = -\tfrac{c}{a} + (\tfrac{b}{2a})^2$$

$$(x + \tfrac{b}{2a})^2 = \frac{b^2 - 4ac}{4a^2}$$

or

$$(x + \tfrac{b}{2a})^2 - \frac{b^2 - 4ac}{4a^2} = 0$$

Now the equation is written in the vertex form where $(-\tfrac{b}{2a})$ represents the x-coordinate of the vertex, and $-\frac{b^2 - 4ac}{4a^2}$ represents the y-coordinate of the vertex. If you take the square root of both sides of the previous step to solve for x, this leads you to the quadratic formula, which can be used to solve all quadratic equations.

$$\sqrt{(x + \tfrac{b}{2a})^2} = \sqrt{\frac{b^2 - 4ac}{4a^2}}$$

$$x + \tfrac{b}{2a} = \pm\frac{\sqrt{b^2 - 4ac}}{2a}$$

$$x = -\tfrac{b}{2a} \pm \frac{\sqrt{b^2 - 4ac}}{2a}$$

or

$$x = \frac{-b \pm \sqrt{b^2 - 4ac}}{2a}$$

This links the quadratic formula to the process of completing the square.

The Classroom

Now that you have explored your own thinking about solving equations, you will explore what this looks like in the classroom. How do students develop their ability to make sense of tasks and use tools to solve mathematical problems? What experiences do students gain in middle school in algebraic contexts that can be further developed throughout the high school curriculum? Videos provide useful windows into students' thinking. In the first video, students explore a bridge problem similar to the one in figure 2.5 and examine the number of beams involved in creating bridges of different lengths. Take time to carefully watch this video in its entirety.

www.solution-tree.com/Making_Sense_of_Models

After seeing the students and the teacher interact with this task, what initial reactions did you have? What did you notice about the role of the teacher? How did the teacher facilitate the conversation so that students needed to make sense of each other's reasoning?

How did students respond to the question regarding how many beams there would be in a bridge of length 5 and a bridge of length 10? You may recall that one student uses a table as a mathematical tool to analyze the task and states, "I predicted that each time, it is just adding four to the beam." While being able to make sense of the problem through the use of a table is important, there are greater mathematical goals on the horizon. The teacher asks, "How could you figure out a bridge of length ten without writing the whole table to get to ten?" Some students use toothpicks, others use drawings, and some use a table of values as tools to think through the mathematics. The teacher does not explicitly give these strategies to the students, nor does she give any hint as to which tool may be an appropriate one to use. In this video, the students are engaged in Mathematical Practice 5, "Use appropriate tools strategically." Based on the context and the information provided, students decide which tools are appropriate and then strategically employ them to analyze the bridge problem.

Gradually students need to develop the ability to reason numerically and algebraically to make sense of expressions and algebraic equations that extend well beyond the specific values in a finite table. Students can use graphs of algebraic functions to reason with equations and analyze potential solutions to equations and make sense of representations in different ways.

In the second video, high school students make sense of algebraic and graphic solutions of the equation $4x + 2 = 2x + 4$ and then determine how the solutions are related.

www.solution-tree.com/Linking_Graphical_and_Algebraic_Solutions_to_Equations

After viewing this video, what are your thoughts? How was instructional time spent during the lesson? How did the teacher respond to students' ideas? The teacher begins by asking the first group how it is thinking about the task mathematically. One student shares an insight that the algebraic solution is related to the graphic solution when the value of *x* is equal to 1. This provides an opportunity for the teacher to engage the other group members in the task. Notice that instead of asking the student to provide an explanation to this, the teacher encourages the group members to make sense of each other's solutions and ask questions. Does the teacher tell the students how to solve the task? Does she tell the students to consider each side of the equation as a function and graph it to solve the equation? She is purposeful in the way she facilitates the discussion by posing questions and providing ample time for the students to explore the task. Through discussion, the students explore the task and realize that the two lines correspond to the graphs of the function determined by each side of the equation.

TQE Process

At this point, it may be helpful to watch the second video again (page 53) and pay close attention to the task, questioning, and opportunities to collect evidence of student learning.

The TQE process can help you frame your observations. Teachers who have a deep understanding of the mathematics they teach:

- Select appropriate *tasks* to support identified learning goals
- Facilitate productive *questioning* during instruction to engage students in Mathematical Practices
- Collect and use student *evidence* in the formative assessment process during instruction

The *task* chosen for this lesson supports students to make sense of algebraic and graphical solutions of an equation by connecting different representations. Some students prefer to solve the task algebraically, and others prefer to work from a graphical representation. This task is purposefully designed to provide the opportunity for students to explore how different forms of the solution process are related.

The teacher uses *questioning* to gather information regarding how students are making sense of the problem. The teacher asks questions like "What are you doing?" and "How are you thinking about this problem?" to encourage students to explain their thinking. She asks questions of other students in the small groups to encourage making sense of each other's work. Consider the case when one student answers for the group. What does the teacher do to facilitate the discussion among all the members of the group? The teacher asks additional questions to hold all members of the group accountable for the talk and to facilitate the engagement of various individuals in the small groups in solving the task. Note the opportunities the teacher creates when she provides feedback regarding student responses. Purposefully and quite intentionally, she does not indicate which response is correct. This supports the students needing to provide mathematical explanations and justifications to help determine an acceptable solution. This process does not solely rely on the teacher for validation of the mathematics.

With which Mathematical Practices are the students engaged? The emphasis is on making mathematical arguments that rely on explanations and justifications based on mathematically valid points. Students are expected to provide justifications and also to examine each other's reasoning very closely and analytically. Hence the students are engaged in Mathematical Practice 3, "Construct viable arguments and critique the reasoning of others." Engagement in Mathematical Practice 3 enables students to provide sound reasoning in the mathematics classroom and well beyond.

As the teacher is circulating among different groups and asking questions, she is gathering *evidence* of their learning. How does the teacher choose questions to learn about the different levels of student understanding of the task? Asking questions like "What did he mean by that?" or "What did she say?" encourages students to ask questions and holds students accountable in order to make sense of each other's solutions and thinking. Not only is the student who is providing the strategy accountable for providing the explanation, but the other individuals are also responsible for listening, processing, and making sense of the mathematics involved in exploring the task. This allows the teacher to learn about each student's thinking. This information provides her with formative assessment data to use in thinking about how to support individual students, small groups of students, and the whole class.

The Response

In high school mathematics, the focus of instruction shifts to include the use and connection of graphic and algebraic techniques to understand the meaning of solving equations. Students are expected to illustrate a deep understanding of why mathematical methods work as they respond to rigorous tasks. Since traditional classroom instruction focuses on manipulating algebraic expressions, students develop proficiency in simplifying and manipulating expressions to solve equations. When the focus shifts to understanding, students will demonstrate new misconceptions, such as not being able to explain what a solution represents graphically or what a solution means in a real-world context. For instance, students may be asked to determine the solution to a system of linear equations. Students may be able to determine the coordinate point at which the lines intersect, but without the development of meaning through understanding multiple representations, they may not be able to explain the meaning of the point as a solution to the system of equations. They may not understand how to connect that meaning to the context of the problem. Here, an approach that involves connecting the equations to the context can support students being able to explain the meaning of their solution.

What docs it mean to solve a system of equations in a real-world context? While the ability to solve problems through algebraic procedures can be very useful, teaching how to solve mathematical tasks strictly in an algebraic sense limits students' abilities to explain solutions and situations that require understanding multiple representations. To develop a deeper and more thorough understanding of mathematical concepts, it is important that students are given opportunities to engage in rich tasks that challenge them to consider multiple representations and flexibly transition from one representation to another. Such flexibility in understanding becomes increasingly useful as students progress through the high school curriculum and beyond and allows for multiple entry points into solving challenging contextualized tasks.

Reflections

1. What do you feel are the key points in this chapter?
2. What challenges might you face when implementing the key ideas from this chapter? How will you overcome them?
3. What are the important features for developing an understanding of the structure of equations, and how will you ensure your instruction embeds the support needed for these features?
4. Select a recent lesson you have taught or observed focused on the structure of equations. Relate this lesson to the TQE process.
5. What changes will you make to your planning and instruction based on what you read and considered from this chapter?

CHAPTER 3

Geometry

Geometry often provides the first opportunity for a formal exploration of proof. Students make sense of and apply both inductive and deductive reasoning in this exploration. Inductive proof could be thought of as connecting to recursive thinking in the study of arithmetic sequences in algebra, while deductive thinking links more closely to determining explicit rules for those sequences. Thinking back to the bridges problem from chapter 2 (see figure 2.5, page 42), the *recursive rule* would be that the number of beams in the next bridge is equal to the number of beams in the current bridge with four additional beams added. The number of beams in each subsequent bridge is based on the previous example. An *explicit rule* for the number of beams in any bridge is not based on the number of beams in a prior bridge but is rather determined using the length of the bridge.

Specifically, *inductive reasoning* is using observations and examples to reach a conclusion. For example, when you recognize a pattern in a number of different isosceles triangles that the base angles are congruent, you may inductively conclude that base angles of isosceles triangles are congruent. However, this type of reasoning is always subject to finding an example that does not follow the pattern. For example, in the bridges problem if you only examine a bridge of length 4 (see figure 2.6, page 43), you might conclude that a rule for the number of beams in a bridge of any length is $3 + n(n - 1)$. However, upon further examination, you would see that this rule cannot be generalized to bridges of other lengths.

In geometry, *deductive reasoning* is using definitions, axioms, and properties in a logical argument to reach a conclusion. Building on the definition of congruence, students construct proofs regarding congruent shapes and conclude that corresponding parts of congruent shapes are also congruent. In the bridges problem, as students explore a rule for the number of beams in a bridge of any length, they may construct an expression that represents the number of beams used for any length bridge without the need for determining the lengths of bridges preceding a targeted length. By seeing the relationship between bridge lengths and using the structure of the bridge to define that relationship, students may reason that the length of any bridge can be represented as $4n - 1$, where n represents the number of beams. Learning how to reason about geometric relationships provides a basis for thinking about problem solving. It is important for instruction to demonstrate how proof can play a role in solving real-world problems.

Additionally, the study of geometry includes the development of a sense of space and visualization—such as with using transformations—as well as the relationships among the measurements of the length, area, and volume of objects. Geometry provides a bridge between many topics in mathematics. It connects functions to their representations, proportions to similar triangles, and triangles to trigonometry.

The Challenge

One aspect of the study of geometry involves building understanding of trigonometric ratios. In investigating trigonometric ratios, you are linking concepts of similarity, right triangles, and proportional reasoning. This investigation leads to later work with the unit circle and trigonometric identities. How do you connect similarity with ratios of the lengths of sides of right triangles (see figure 3.1)?

DO NOW

ΔABC is a right triangle. The length of a is 4.5 units, b is 6 units, and c is 7.5 units.

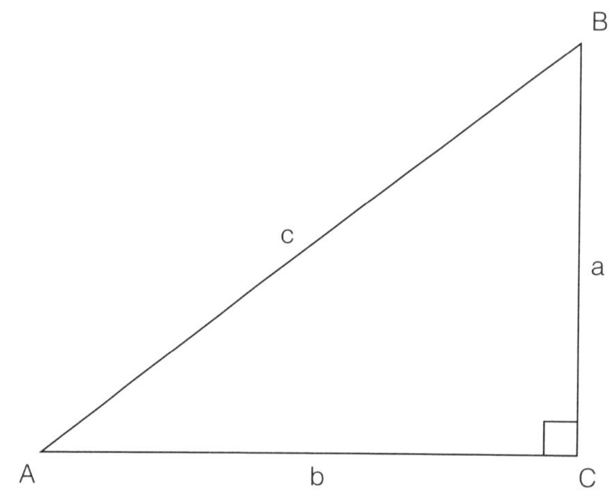

Create a similar triangle to ΔABC with a hypotenuse of length 1. What are the lengths of the remaining sides? How do these lengths relate to the values of sine and cosine for angles A and B in ΔABC?

Figure 3.1: Trigonometric ratios task.

In ΔABC, $\sin A = \frac{4.5}{7.5} = 0.6$ because the sine function is defined to be the ratio of the opposite side from the angle to the hypotenuse. What was the ratio for the similar triangle you created? The ratio should also be 0.6. Students know from the middle grades that this ratio remains constant in similar triangles. In the similar right triangle you created, the hypotenuse is one unit, so the ratio could be represented as $\frac{0.6}{1}$; therefore, the length of the opposite side represents the value of the sine function for $\angle A$. This is the same for the cosine ratio in these two triangles. The length of the adjacent side is the same as the value of the cosine function. Using a similar triangle with a hypotenuse of length 1 provides direct representation of the values of the sine and cosine functions for the complementary angles in the right triangle (see figure 3.2).

What do you notice about the relationship between $\sin D$ and $\cos E$? Think about how the trigonometric functions of sine and cosine and the sides of the triangle relate to one another. Since both $\sin D$ and $\cos E$ are determined by the ratio $\frac{EF}{DE}$, you know that they must have the same value. This is due to $\angle D$ and $\angle E$ being complementary angles. This makes sense when you think of the origin of the word *cosine*. The cosine function can be thought of as the sine of the complementary angle in the right triangle, hence "complementary sine," or cosine.

In determining trigonometric ratios for a right triangle with a hypotenuse of length 1, you are setting the stage for the development of the unit circle. Figure 3.3 provides an illustration of how the unit circle is

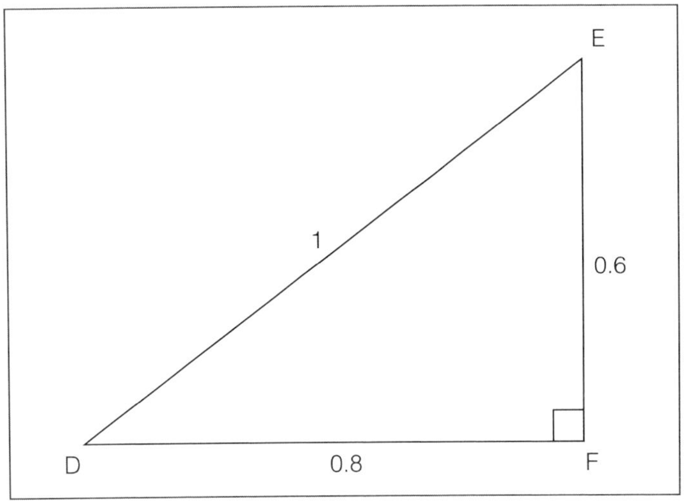

Figure 3.2: Δ*DEF* is similar to Δ*ABC* with a hypotenuse of length 1.

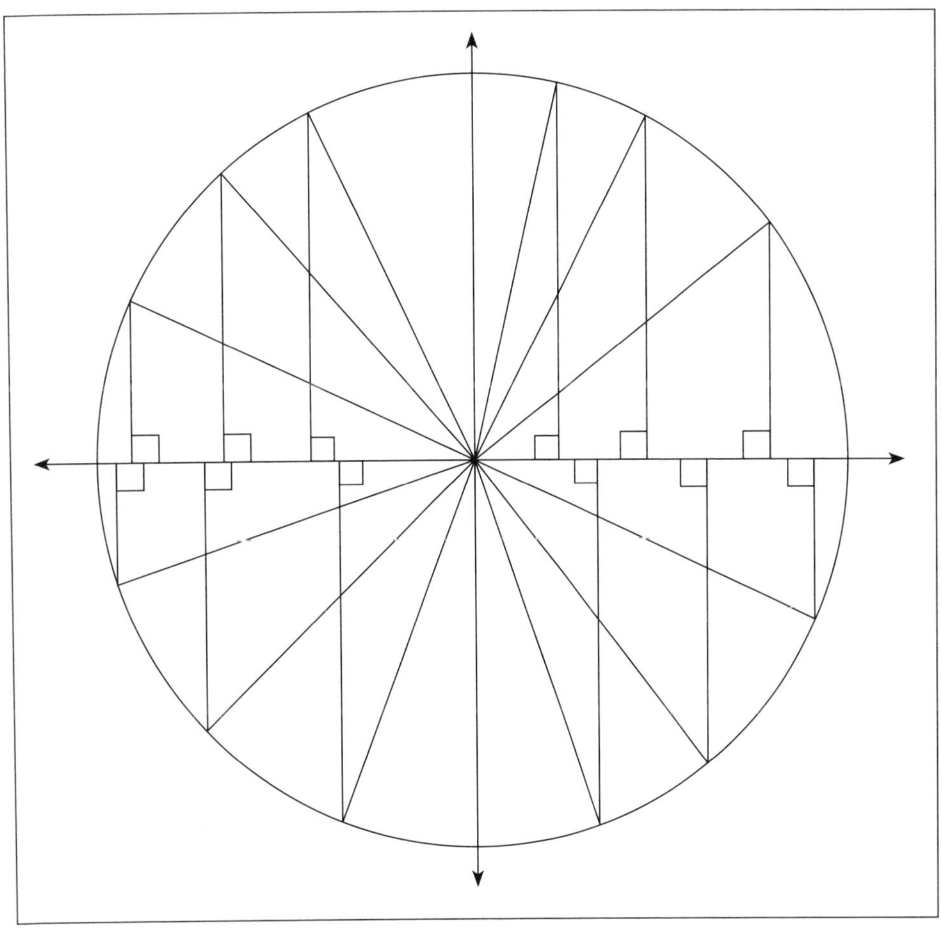

Figure 3.3: Multiple right triangles with hypotenuse equal to 1 form the unit circle.

generated from right triangles, each having a hypotenuse with a length of one unit. Notice how one of the vertices of the non-right angles in each triangle serves as the origin of a coordinate plane and is concurrent for all of the triangles. The vertices of the other non-right angles form a circle of radius 1—the unit circle.

How does figure 3.3 connect to the Pythagorean theorem? The Pythagorean theorem states that in a right triangle, $a^2 + b^2 = c^2$, where a and b are the legs of the right angle and c is the hypotenuse. As previously demonstrated, the side lengths a and b of the right triangles forming the unit circle can also be represented by trigonometric functions, so another way of writing the Pythagorean theorem is $(\sin A)^2 + (\cos A)^2 = 1$. This is an important trigonometric identity that will be explored in later mathematical study. How does this connect to the equation of a circle? As these are also values of x-coordinates and y-coordinates on the coordinate grid, you can extend this reasoning to show that $x^2 + y^2 = 1$, the equation of the unit circle.

These extensions provide opportunities to make sense of trigonometric functions in ways that set the stage for studying algebraic representations of circles and applications of the unit circle.

The Progression

Students have reasoned with shapes since kindergarten by examining attributes and creating increasingly precise definitions. They learned about angles, parallel lines, and right angles in grade 4 and the coordinate plane in grades 5 and 6. In the middle grades, students explored the relationship between surface area and volume of three-dimensional shapes; learned about all four transformations, congruence, and similarity; and proved and applied the Pythagorean theorem. Each of these areas is extended in high school. Following is a learning progression for how students extend this knowledge in studying geometry in high school.

- Use transformations to demonstrate geometric properties.
- Understand congruence and similarity in terms of transformations.
- Prove geometric relationships involving lines, angles, and geometric figures.
- Connect geometric descriptions and relationships to algebraic justifications, including coordinate geometry.
- Use trigonometric ratios and the Pythagorean theorem to solve problems.
- Connect relationships in two- and three-dimensional figures.

Students need to be provided many different experiences in order to make sense of this learning progression and to make connections within geometry and between algebraic and geometric concepts. Instruction needs to support students to reason through the use of multiple representations, explanation, and proof. These experiences should also develop conceptual understanding that makes sense of procedures and processes.

Grade 8

Students use translations, rotations, and reflections to demonstrate understanding of congruence of geometric figures. Figures are shown to be congruent when one shape can be exactly superimposed onto another through rigid motion transformations. The use of dilation, a non-rigid motion transformation, supports the development of understanding similarity. The Pythagorean theorem is proved and applied to solve problems in the Cartesian coordinate plane and in the real world. Students solve mathematical and real-world problems involving volume.

High School

Students build on their understanding of translations, rotations, reflections, and dilations to develop richer understandings of congruence and similarity. Definitions of transformations are extended and described with precision. Transformations are also used to support the development of proving congruent and similar relationships between figures. Students represent their thinking geometrically through constructions and connect to algebraic concepts through coordinate representations using function notation.

Students relate properties of geometric figures to the coordinate grid and algebraic representations. These connections are used to create geometric figures such as parabolas and circles as well as to prove theorems involving geometric figures. Circle relationships are used to make connections among chords, angles, area, and sectors.

Students build on their work with right triangles and the Pythagorean theorem from grade 8 as they explore trigonometric relationships. Trigonometry is the study of the relationships of angle measures and lengths of sides of triangles. This connects student understanding of similarity to relationships within right triangles. Students solve problems involving trigonometry both in and out of context.

The Mathematics

Making sense of shape and space provides the tools to describe the world. Students in the middle grades are introduced to transformations, congruence, and similarity. High school students formalize these understandings, using them to explore transformations and demonstrate geometric relationships. Students will also use proof and constructions to investigate congruence and similarity.

Exploring Transformations

One important aspect of the study of geometry in high school entails establishing and extending understanding of transformations and how they relate to other mathematical topics. In considering rigid motion transformations, you are connecting the impact of reflections, rotations, and translations on sides and angles of figures as well as the consequences of performing each type of rigid motion transformation to the position and orientation of these figures. The following investigation connects similarity, non-rigid transformations, proportional reasoning, trigonometric relationships, and function representations (see figure 3.4, page 62).

How can transformations be used to map a pre-image to an image that is similar but not congruent? Perhaps you started with a translation of 5 units to the right followed by another translation—down by 1 unit—resulting in a triangle similar to ΔDEF with angle A'' concurrent with angle D from figure 3.4 (see figure 3.5, page 62).

DO NOW

ΔDEF is the image of ΔABC following a number of transformations.

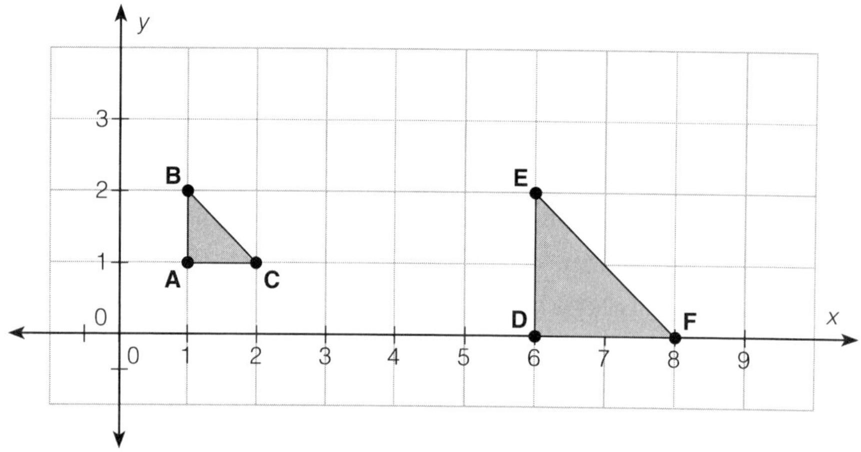

Determine a series of transformations that would map ΔABC to ΔDEF.

Figure 3.4: Geometry transformation task.

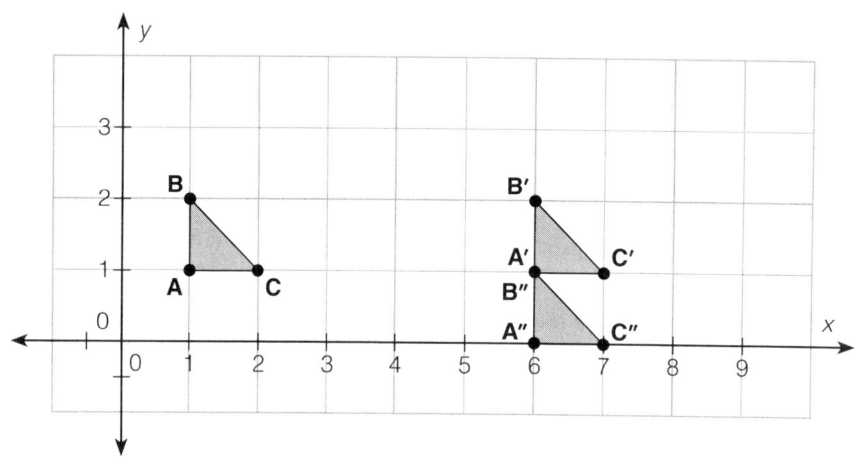

Figure 3.5: A series of two translations.

A helpful discussion for students is to justify how they know that $\Delta A''B''C''$ is congruent to ΔABC. They should deduce that since all of the vertices are moved in the same direction and by the same number of units, the new triangle must be the same, or congruent, because a triangle is determined by its vertices. However, ΔDEF is not congruent to ΔABC. What transformation will map $\Delta A''B''C''$ to ΔDEF? Dilations map pre-images to images that are similar but not congruent. The scale factor determines the expansion or contraction of the pre-image about the center of dilation. If the center of dilation is located on the pre-image, then that position in space is conserved. For example, by performing a dilation on $\Delta A''B''C''$ with the center of dilation at A'' and a scale factor of 2, the image, $\Delta A'''B'''C'''$ has a vertex A''' that is concurrent with A'', and the ratio of the length of the sides of $\Delta A'''B'''C'''$ to the length of the sides of $\Delta A''B''C''$ is 2:1 (see figure 3.6). The vertices of $\Delta A'''B'''C'''$ concur with the vertices of ΔDEF. If you used this sequence of transformations, you performed the following: (1) a horizontal translation, (2) a vertical translation, and lastly, (3) a dilation.

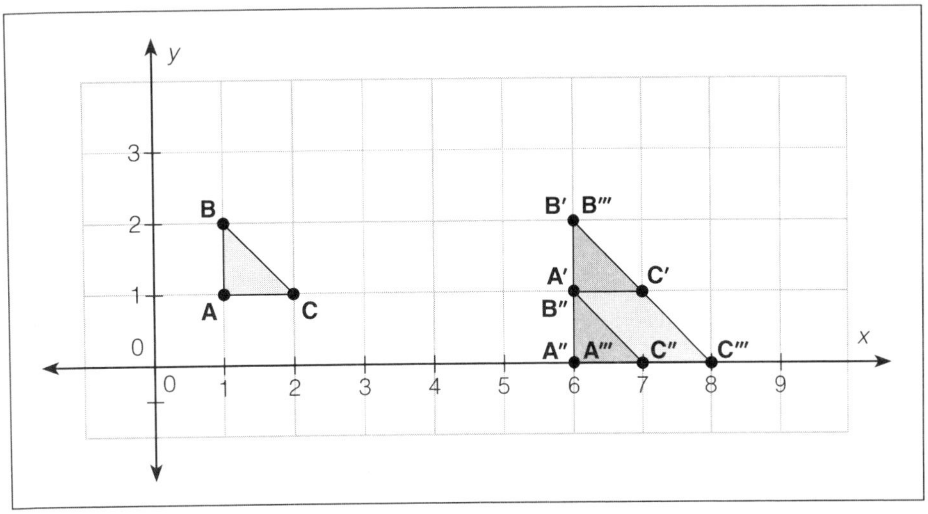

Figure 3.6: Modeling the transformation of ΔABC to ΔDEF.

You might have performed a dilation and then just one translation to map ΔABC to ΔDEF. If the center of dilation is located at A and the scale factor is 2, how might the remaining mapping be accomplished with just one translation (see figure 3.7)?

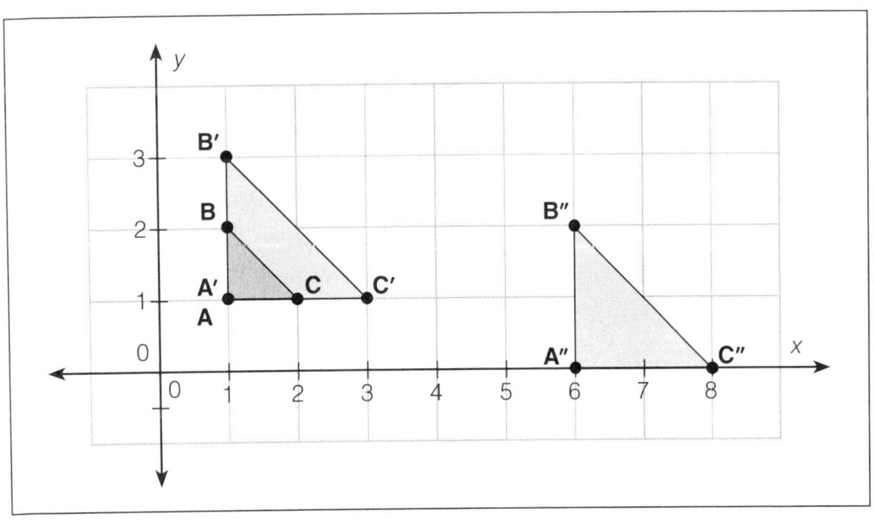

Figure 3.7: A dilation then a translation.

A vector can describe a combination of horizontal and vertical translations, such as five units to the right and one unit down. This can be represented using function notation such as $f(x + 5, y - 1) = g(x, y)$, where $f(x, y)$ is the pre-image (in this case, $\Delta A'B'C'$) and $g(x, y)$ is the image ($\Delta A''B''C''$, or ΔDEF). Students should be encouraged to explore both types of translations to support later work with functions.

In what other ways could the mapping have been accomplished? When performing geometric transformations, you can compose the geometric moves in a variety of ways to produce the desired results. Consider the transformations portrayed in figure 3.8.

DO NOW

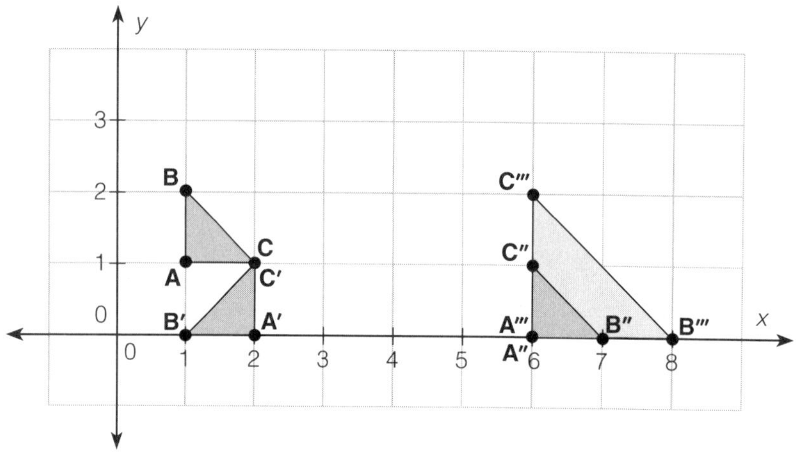

Describe the sequence of transformations that map ΔABC to $\Delta A'''B'''C'''$.

Figure 3.8: A sequence of transformations.

The first transformation is a rotation about center C, but what is the angle of rotation? It can be described in more than one way. You can think of an analog clock and describe the rotation as a 90° rotation in the counterclockwise direction or a 270° rotation in the clockwise direction. You can also describe the rotation using positive and negative angles. Students are often surprised to learn that positive angles are described by counterclockwise rotations. This will be addressed when students explore trigonometry with the unit circle.

What transformation maps $\Delta A'B'C'$ to $\Delta A''B''C''$ in figure 3.8? It cannot be a translation because the orientations of the pre-image and the image are not the same. It is a reflection. This type of transformation is determined by a line of reflection. The line of reflection must be equidistant from corresponding points in the pre-image and the image. In this case, the line of reflection is described as the line $x = 4$. The final transformation is a dilation and is similar to the dilation described earlier as the center and scale factor are the same (see figure 3.6, page 63). Is $\Delta A''B''C''$ the same as ΔDEF? However, in this dilation, B''' maps to F, and in the previous dilation, B''' maps to E. What does this tell you about ΔABC and ΔDEF? The only way that these two sequences of transformations could result in congruent shapes is if both triangles have two congruent angles, as is the case with these angles. Because these triangles are isosceles, it is not immediately obvious that the sequence of transformations mapping ΔABC to ΔDEF is not complete. There needs to be a reflection of $\Delta A'''B'''C'''$ in order to map A to D, B to E, and C to F as is required with pre-images and images.

How might students determine the equation for the line of reflection? They might use two points to determine the slope of the line and then use the point-slope formula. What points might they use? Since this is an isosceles triangle, the line of reflection will contain A''' and the midpoint between B''' and C'''. The slope of the line containing these two points is $m = (1 - 0)/(7 - 6)$, or 1. They might also draw the line of reflection and then see that the slope is 1 and indicate that the equation for the line is $y - 0 = 1(x - 6)$ or $y = x - 6$ (see figure 3.9). Asking students to justify their solution process helps them understand mathematics as a coherent subject in which they use what they learn in previous years or with previous topics to make sense of new ones.

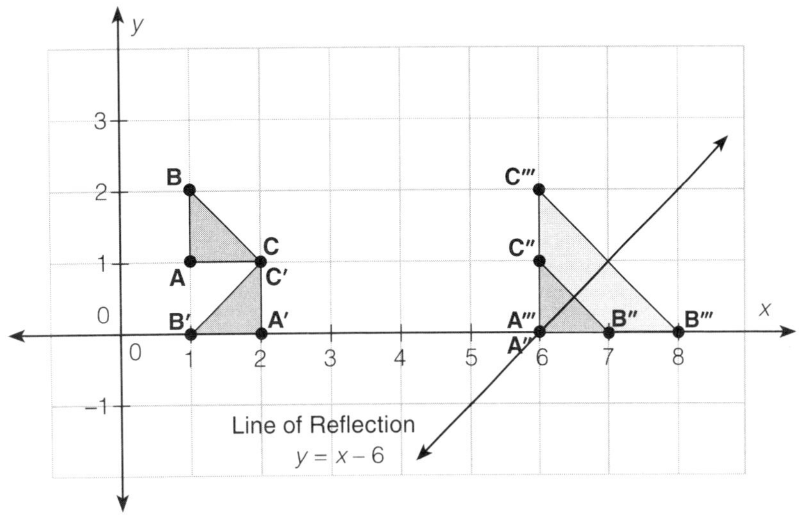

Describe the sequence of transformations that will map ΔABC to ΔDEF from figure 3.4.

Figure 3.9: Completing the transformation of ΔABC to ΔDEF.

It is important for you to consider how the tasks you use with your students help in learning not just your course's content, but also the content on the horizon. Students use transformations to make sense of graphical representations of functions. Experiences like the one provided in the geometry task help students use transformations and parent functions to graph other functions. For example, $y = x$ is a parent function for the lines that contain $\overline{DF}$ and $\overline{JL}$ (see figure 3.10, page 66). How is the function transformation for the line that contains $\overline{DF}$ transformed from the parent function $y = x$? The transformation that maps $\overline{AC}$ to $\overline{DF}$ is a vertical translation of four units in the negative direction. Notice that the new function for the line that contains $\overline{DF}$ is $y = x - 4$. This new function indicates a shift of the y-intercept by four units in the negative direction.

How is ΔABC mapped to ΔJKL? It is a reflection about the y-axis. How is this transformation reflected in the transformed parent function for the line that contains $\overline{JL}$? This new function is $y = -x$. What would the function have been if ΔDEF had been reflected over the y-axis? The new function for the line that contains $\overline{D'F'}$ would be $y = -x - 4$. Observe how both transformations are evident in the function. As students begin to notice how changes in parent functions result in changes in graphs of those functions, they are engaging in Mathematical Practice 8, "Look for and express regularity in repeated reasoning."

In addition to supporting connections between algebraic and graphical representations of functions, transformations are useful when exploring congruence and similarity. Students see that if a sequence of rigid motion transformations can map one figure to another, then the two figures are congruent. If the sequence includes a dilation, then the shapes are no longer congruent; however, they are similar.

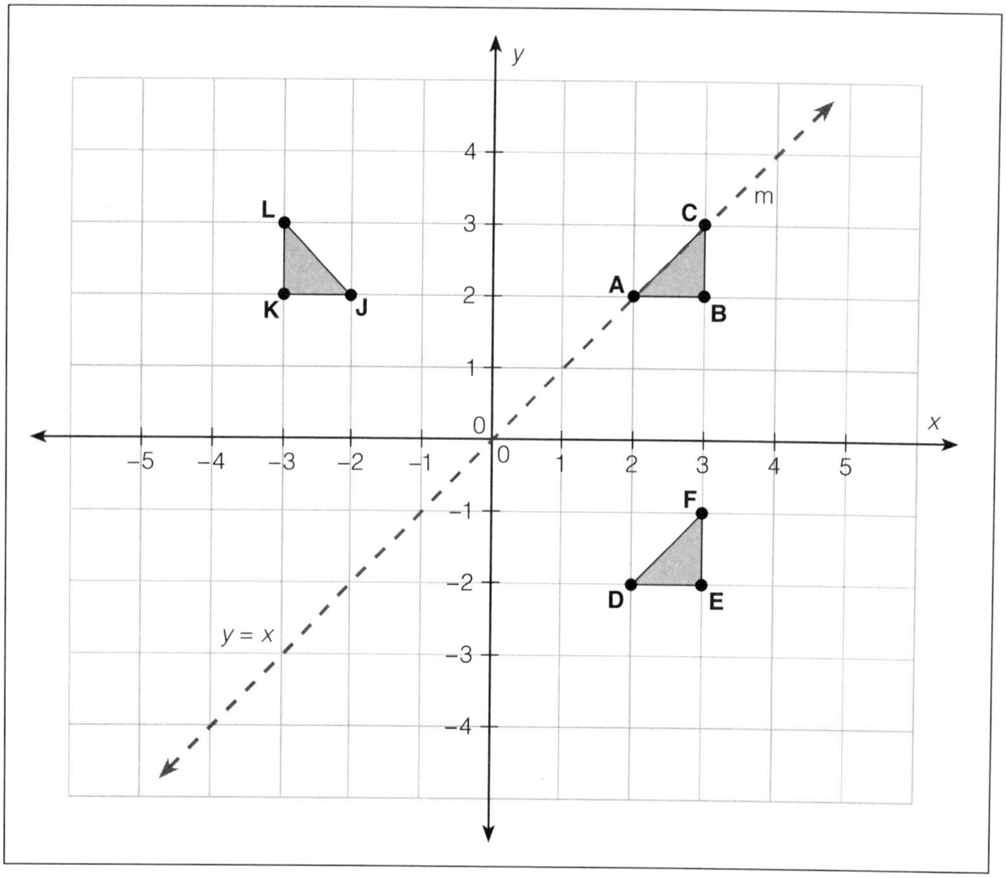

Figure 3.10: Exploration of a parent function.

Exploring Congruence Through Proof

How do you develop an understanding of deductive reasoning in geometry? Consider the task in figure 3.11.

DO NOW

> Joy, a building designer, asks for a triangular building support to be constructed with the following dimensions: ΔABC with $AB = 6$ meters, $AC = 6.4$ meters, and $m\angle BCA = 52°$. Marni builds the triangle based on the given requirements. Joy complains that the triangle is not constructed correctly, but Marni shows that the triangle is built to the correct dimensions. Explain how both Joy and Marni can be correct.

Figure 3.11: Proving triangles congruent task.

What could have happened for two triangles to have three corresponding parts the same yet be different triangles? As you probably know, you do not need to be given the measures of all corresponding angles and sides of two triangles to prove congruence of triangles. This is an important distinction for students as they learn to develop proofs—learning what information is *necessary* and what information is *sufficient* to use congruence theorems. You likely know the different triangle congruence theorems—side-side-side (SSS), side-angle-side (SAS), angle-side-angle (ASA), and angle-angle-side (AAS). Helping students determine how these relationships can be used to create proofs is an important factor in developing reasoning in geometry. Each of the congruence theorems need only three congruent corresponding parts. Since the task in figure 3.11 has three pieces, can you prove the two triangles should be congruent?

You may have decided to start thinking about this task by drawing a triangle that represents the given information (see figure 3.12).

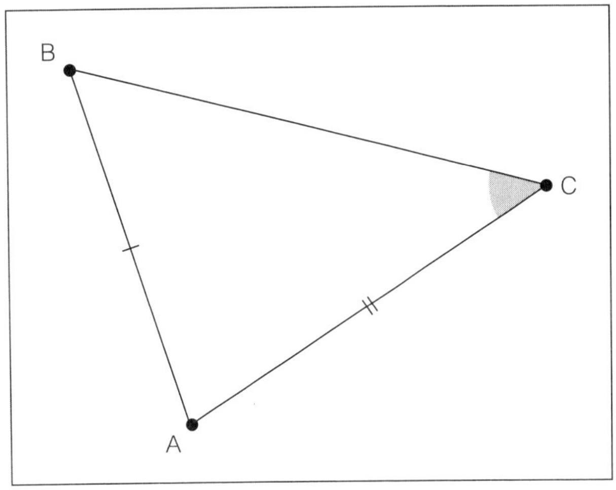

Figure 3.12: Drawing of ΔABC.

This triangle matches the given information but is not the only possible triangle that could be drawn from these dimensions. This task is an example of the side-side-angle (SSA) relationship (where the length of two sides are congruent as is one of the angles that is not composed of the given sides), which is not a congruence theorem. While you may create another triangle congruent to the one in figure 3.12 based on the given information, there is another possibility. Consider how you might build a triangle using the given information. You could start with the angle constructing a 52° angle at point *C* (see figure 3.13).

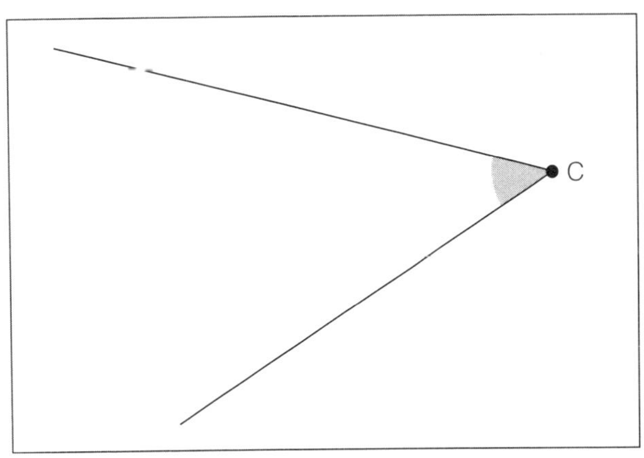

Figure 3.13: Step one of constructing the triangle.

Next, you probably thought about how to include the length of the two known sides. Since you are given $AB = 6$ and $AC = 6.4$, and you currently only have C, then you can measure the length of $\overline{AC}$ (see figure 3.14).

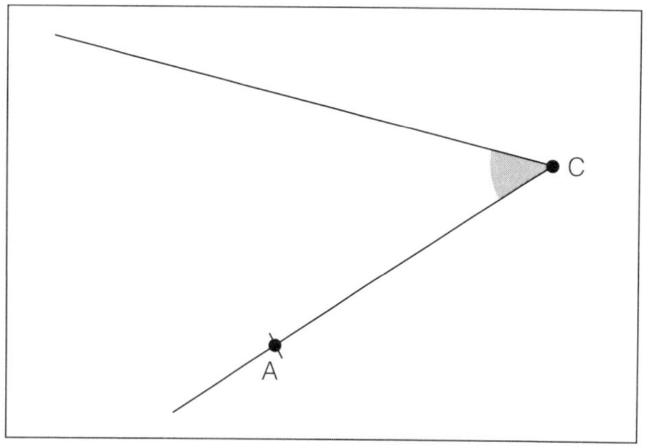

Figure 3.14: Step two of constructing the triangle.

This step gives you two of the three components that are provided regarding the triangle. What still needs to be determined? The remaining component is the length of $\overline{AB}$, which is 6 units. How could you determine the location of point B? If you draw an arc with a center at A and radius of 6 units, you notice that it intersects the other ray in two places (see figure 3.15).

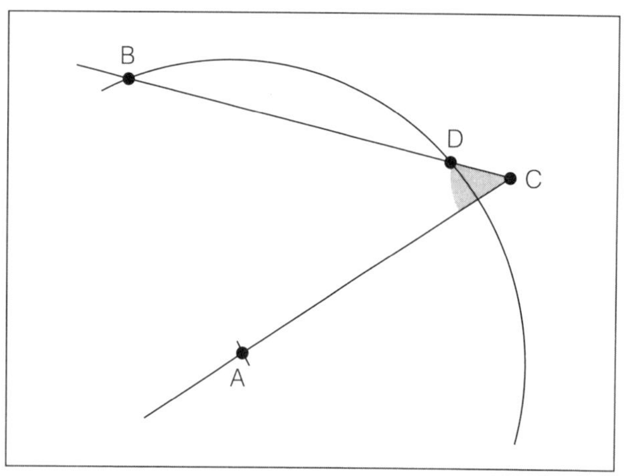

Figure 3.15: Step three of constructing the triangle.

These two points of intersection indicate two possibilities for the location of vertex B. These two ways of representing the triangle are illustrated in figure 3.16 with ΔABC and ΔADC.

So it is possible that Marni built ΔADC while Joy expected ΔABC. This example demonstrates that you need to be careful to consider all possible ways that triangles can be created. Your students may feel that having three pieces of information is sufficient to prove triangles congruent, no matter what pieces of information are given. They may also feel that being able to draw a picture that shows they are the same supports that view. Students often feel that if they can draw a given relationship, then that relationship

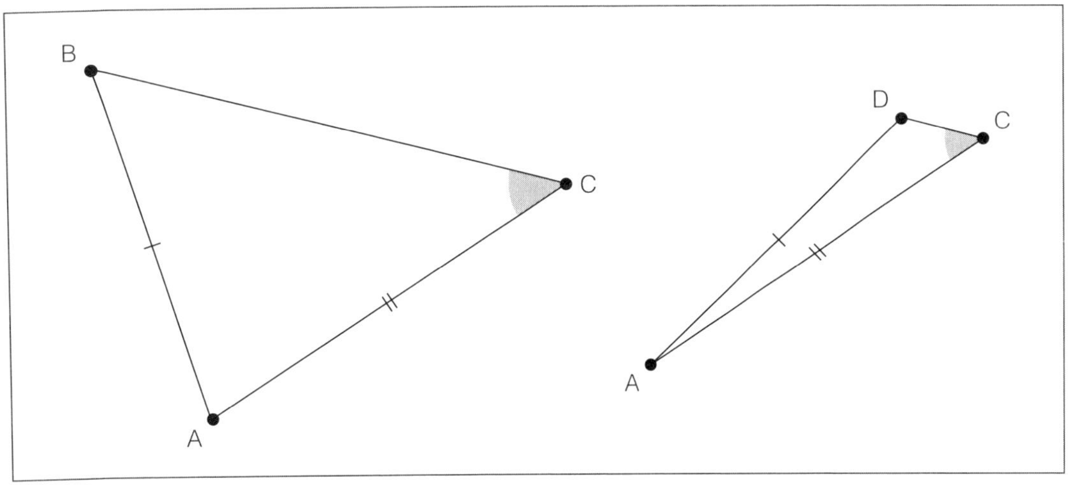

Figure 3.16: Two triangles that meet the conditions of the task in figure 3.11.

is always true. This is a misuse of inductive reasoning. The ability to draw two congruent triangles does not *prove* any two triangles with those given measurements are congruent, such as in the case of SSA. This is an important understanding in investigating triangle congruence and proof—how you need to know that the triangles *must* be the same, not that they *may* be the same. This depth of understanding regarding proof is an important outcome for students in high school mathematics.

Exploring Similarity Through Constructions

Developing understanding of congruence and similarity is fundamental in learning geometry. The trigonometric ratios task (see figure 3.1, page 58) demonstrated one way that concepts in similarity are used by examining geometric relationships through measurement and proportionality. How could you explore these same concepts in context? Consider how similarity is used in the task in figure 3.17.

DO NOW

> You have a picture that is 3 units by 5 units. You want to make a copy of this picture in the same proportion that is three times the area. What are the dimensions of the enlargement? Use constructions and the Pythagorean theorem to draw the enlarged picture.

Figure 3.17: Enlarging a picture task.

What are your initial thoughts? How will you determine the new measurements? You are probably thinking about how increasing linear dimensions impacts the measurement of area. You may be thinking about how you need to multiply both dimensions by the same number in order to ensure similarity, yet still get a threefold increase in area. This is an example of connecting ideas in one- and two-dimensional relationships and is a topic in which misconceptions often occur with students. Many students initially think that to solve this task, they should multiply both dimensions by 3 to get a picture that is 9 units by 15 units. The challenge is when they attempt to demonstrate the relationship between the old and new area, the area has gone from 15 square units to 135 square units. Instead of enlarging the picture by a factor of 3, they have actually increased the area by a factor of 9.

How can you increase the area by a factor of 3 and still keep the new shape similar to the original? Since each dimension is a factor of the area, you need to use the same scale factor on each. Since the desired increase in area is an increase of a factor of 3, you should multiply each dimension by $\sqrt{3}$ in order to maintain similarity and get an area three times as large. This means that you should multiply each of the dimensions by approximately 1.732 (a decimal approximation of $\sqrt{3}$) in order to determine the new dimensions. The new photograph should be approximately 5.196 units by 8.660 units.

But the directions indicated that you need to use constructions and the Pythagorean theorem to draw the enlarged figure. How can you create the measurement of $\sqrt{3}$ using only a compass and straightedge or their representation in a DGE? What do you know that can help you determine a length for $\sqrt{3}$? You can define a unit length and construct a segment that is $\sqrt{3}$ units (see figure 3.18).

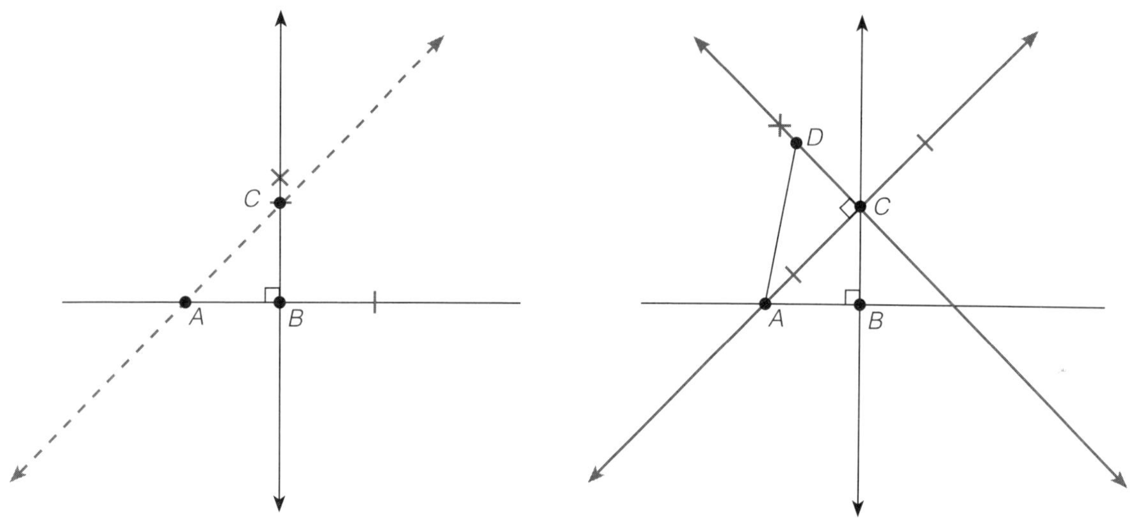

Step 1. Construct a length of $\sqrt{2}$ (segment $\overline{AC}$). (Note: $\overline{AB}$ is defined to be length 1.)

Step 2. Construct a length of $\sqrt{3}$ (segment $\overline{AD}$). (Note: some previous construction lines have been removed to show the highlighted aspects of the discussed relationship.)

Figure 3.18: Constructing a length of $\sqrt{3}$.

As you need an exact measure of the square root of 3, one strategy would be to use your knowledge of the Pythagorean theorem to construct $\sqrt{2}$ and use this length to build the length of $\sqrt{3}$, again applying the Pythagorean theorem. Figure 3.18 shows the construction of a right triangle, which begins with building perpendicular lines, measuring a unit length on each line ($\overline{AB}$ and $\overline{BC}$), and creating $\overline{AC}$, whose length is $\sqrt{2}$. The figure on the right shows the construction of a right triangle with angle ACD as the right angle. With sides 1 (CD) and $\sqrt{2}$ ($\overline{AC}$ from the triangle construction on the left), the hypotenuse measures $\sqrt{3}$.

AD will be the length you use to construct a rectangle that has an area three times the area of a 3-unit by 5-unit photograph, by constructing perpendicular lines and copying segments. The construction of the desired rectangle is shown in figure 3.19.

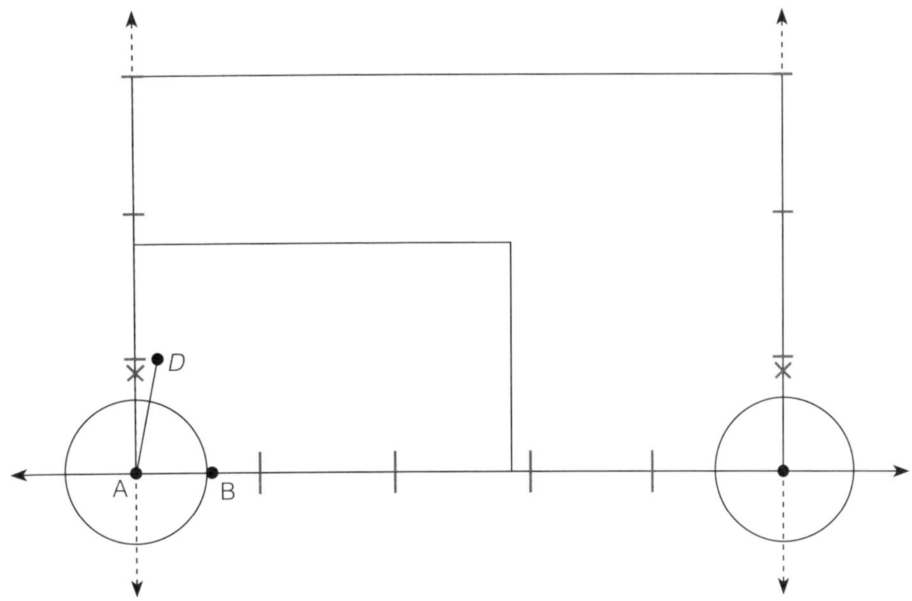

Step 1. Photograph that is 5 units by 3 units, with $\overline{AF} = \sqrt{3}$ units. $\overline{AB}$ is 1 unit.

Step 2. *AJKL* has an area three times the area of *ACDE*.

Figure 3.19: Constructing rectangles whose areas are in the ratio of 1:3.

The approximate dimensions of the enlargement shown in figure 3.19 are 5.196 units by 8.660 units as already given but without the need to estimate the length of $\sqrt{3}$. You need to consider more than one way to solve a given task because when students create and view objects and concepts from multiple perspectives, they develop deep understandings of mathematics. This sort of activity supports engagement in Mathematical Practice 1, "Make sense of problems and persevere in solving them."

The Classroom

The videos for this chapter represent examples of high school students engaged in reasoning about geometry. In the first video, students are challenged to determine the location of the image of a flag after rotation. You are encouraged to watch the video in its entirety before proceeding.

www.solution-tree.com/Visualizing_Rotations

How do the students engage with the task? How does the teacher respond to student ideas? Notice how the teacher begins the lesson with little direction to the students, allowing the students to make sense of what needs to be done to accomplish the rotation. The teacher calls the students back together and asks them to reflect on their experience and share what they found. Again, the students initiate the action, providing different ways that they found to improve their understanding of the rotation.

The focus of the lesson stays on the students, from the choice of materials to solve the task to who is providing the explanation on how to solve the task. The teacher is guiding the conversation, making the decision about which responses the class discusses, but the students provide the ideas. The teacher anticipates student errors and ensures that the class discussion addresses those misconceptions.

In the second video, high school students are considering what information is needed to build a proof to determine the similarity of two triangles. Watch this video before proceeding.

www.solution-tree.com/Exploring_What_Is_Needed_to_Prove_Triangle_Similarity

In this video, students are asked to determine what information they need to prove two triangles similar. Throughout the lesson, the teacher guides the students to think about what information is necessary and sufficient for a proof, with the idea that they may not need every relationship that they can determine. Students consider different options for demonstrating similarity, looking both at sides and at angles. The class continues to explore relationships, and the teacher guides the discussion to use clear and precise language, focusing on Mathematical Practice 6, "Attend to precision." The teacher helps the students consider precise language, supporting the students to make sense of the relationships in the diagram with clarity.

Purposeful questioning during the lesson encourages students to think about different ways to compare corresponding parts of the triangles. The focus is on sense making, with students guided to consider different relationships and different ways in which the corresponding parts could be organized. When students consider different combinations of parts, it is the students who explain their reasoning. This allows for the students to learn from one another, and the expectation is for students to attend to each other in order to make sense of each other's thinking. This is an important aspect of effective learning and clear communication.

TQE Process

At this point, it may be helpful to watch the first video (page 71) again. Pay close attention to the task, questioning, and opportunities to collect evidence of student learning.

The TQE process can help you frame your observations. Teachers who have a deep understanding of the mathematics they teach:

- Select appropriate *tasks* to support identified learning goals
- Facilitate productive *questioning* during instruction to engage students in Mathematical Practices
- Collect and use student *evidence* in the formative assessment process during instruction

The *task* for this lesson is for students to conjecture regarding the location of the image of a flag following a rotation. The learning goal is for students to understand that the size, orientation, and distance from the center of rotation are all conserved with this rigid motion transformation. The task supports this learning goal by providing students opportunities to make individual conjectures regarding each of these aspects of rotation and then to test those conjectures prior to discussing the results with classmates. By making the conjectures prior to comparing and discussing them, students are more likely to experience cognitive dissonance if they do not maintain the properties of rotation regarding maintenance of size, orientation, and distance from the center. This aspect of the task is important to the learning goal.

The *questioning* that the teacher uses helps create the dissonance and also resolves it. One of the teacher moves evident in this video is the expectation that students explain their thinking and be ready to explain the thinking of others. This is an expectation at the individual, small-group, and whole-class levels. When the expectation is for students to listen to each other, and to explain each other's thinking, the focus is on Mathematical Practice 3, "Construct viable arguments and critique the reasoning of others." When the teacher asks, "Does it make sense? Is this a rotation of forty-five degrees about the center?" the question is directed to one student. However, both that student's group and the entire class understand that the question may be directed to them, so they all need to be thinking about a response. This level of thinking, and subsequent productive talk, is an expectation that all classrooms need. Productive student discourse comes from constant engagement by the students, and this climate of discourse is fostered by the teacher. Notice how the teacher sometimes asks multiple questions of the same student and sometimes directs questions to another student. This strategy keeps students engaged in the lesson as they follow the thinking of other students in the classroom.

How does the teacher use *evidence* of student understanding to guide her thinking during the lesson? When she asks the students what they noticed, she is finding out what the students are thinking. This is also a way for students to hear what their peers are considering in the lesson. The teacher continues to ask for student contributions until she hears a common error—that the image should touch the point that identifies the center of rotation. When the student identifies that the base of the flag should touch the center of rotation, the teacher turns to the class and asks for its thinking. She asks questions and provides a drawing to help the students communicate their thinking. The teacher is digging into student reasoning and helping both the individual student and the whole class interpret what is being shared. This is an important aspect of the formative assessment process, ensuring that you understand what the student is thinking and you are helping each student communicate that thinking. The teacher also asks the class a number of questions and collects responses, indicating when the students are not in agreement. This lack of agreement provides evidence that a misconception is held by members of the class. These disagreements become investigations for the class. The teacher is then able to target the misconception based on the evidence provided and address it.

The Response

Some students struggle with the development of justifying geometric theorems and relationships through the structure of proof. How do students learn to construct proofs? Students often believe that if they can draw an example of a relationship, then that relationship is true. They are, in essence, proving by example. This sort of thinking is consistent with inductive reasoning and is an important aspect of helping students understand what is to be proven and often to convince them that the proof exists. However, it is not sufficient.

What experiences do you provide to help students in understanding the manner in which proof goes beyond the individual example and must apply to every possible extent that meets the given conditions? They need experience with a wide range of counterexamples and multiple explorations that encourage thinking that looks for unusual ways to put together pieces of information. Later, in chapter 6, you will read about outliers and the role that they play in statistics. Here, you are looking to encourage students to think about outliers, ways that are different than the ordinary and traditional, ways that help them think about perspectives from which they might examine a situation.

While real-world examples will provide a foundation, students will also need examples that help them understand the role of givens, theorems, and properties that lead to making conclusions. Students should have practice organizing the order in which a proof can be structured. This can be accomplished by giving students all of the statements and reasons in a proof and allowing them to make sense of the order of the pieces of the proof. Scaffolding the creation of different proofs will help students build understanding of the deductive logic of proof and is a key piece of geometric reasoning in high school mathematics.

Geometry offers students the opportunity to link visual and abstract mathematical representations together. Strategies developed in earlier mathematics, from elementary, middle, and early high school, can be applied in increasingly sophisticated ways. It is important that, no matter how the coursework is structured, concepts in algebra and geometry are connected together.

Reflections

1. What do you feel are the key points in this chapter?
2. What challenges might you face when implementing the key ideas from this chapter? How will you overcome them?
3. What are the important features for developing an understanding of geometry, and how will you ensure your instruction embeds the support needed for these features?
4. Select a recent lesson you have taught or observed focused on geometry. Relate this lesson to the TQE process.
5. What changes will you make to your planning and instruction based on what you read and considered from this chapter?

CHAPTER 4

Types of Functions

How does the understanding of functions evolve as students experience different aspects of high school mathematics? As students transition from early coursework in high school mathematics to more advanced courses, they use their understanding of linear and quadratic functions (see chapters 1 and 2) to make sense of exponential, logarithmic, and rational functions, among others. As with earlier work with functions, these more complex functions are explored contextually, numerically, graphically, and algebraically. This chapter will focus on rate of change and other characteristics of various function types. Furthermore, the role of parameters in function transformations and the connections between functions and their inverses are examined, while highlighting the role of symmetry and geometric relationships. In chapter 5, tasks that use these mathematical concepts will be explored with a focus on modeling.

The Challenge

The introduction of exponential, logarithmic, and rational functions allows students more opportunities to make sense of the world around them. Different functions describe different phenomena. What are some ways that you can use context to compare functions? Take a moment to consider the task in figure 4.1.

DO NOW

The functions below represent the population growth in four different cities over the past several years. The variable x measures the time in years, with $x = 0$ corresponding to the year 2010. The function values indicate the population growth each year in thousands of people.

Which function describes the city with the greater population growth?

$f(x) = 2x$

$g(x) = x + 2$

$h(x) = x^2$

$l(x) = 2^x$

Figure 4.1: Comparing population growth through function types task.

How did you begin to compare the population growth of these cities? How did you choose which x-values to consider? How did you organize the values you chose? The particular values in the domain you select have important implications for your analysis, as you need to select values that are important to both the functions involved and the context you are investigating. Selecting values that are important to the functions is critical in order to make a proper analysis of the behavior of the functions. In addition, the context provides a real-world connection that allows you to make sense of the mathematics as you explore this task.

You may have noticed that the first two functions given in the task in figure 4.1 represent linear functions, $f(x) = 2x$ and $g(x) = x + 2$. You probably realized that for some values of x, $f(x)$ will be greater, and for other values, $g(x)$ will be greater, but how does this impact the growth of the population? How would you determine which linear function represents the greater population growth? What information would you use to make that decision? You may have identified $f(x) = 2x$ as having a slope of 2 and a y-intercept of 0. Similarly, $g(x) = x + 2$ has a slope of 1 and a y-intercept of 2. How relevant is this information in answering the question posed regarding which of these two functions describes the city with the greater population growth, $f(x) = 2x$ or $g(x) = x + 2$? Would the statement that "the function that has the slope of 2 has greater y-values than the function having a slope of 1" *always* be true, *sometimes* be true, or *never* be true? The slope of the function relates to the growth of the population, so $f(x)$ has a greater growth rate than $g(x)$. In addition, students may know that a function with a greater slope will produce greater changes in y-values; however, it is important to take the y-intercept into account when making the consideration of which function will produce greater y-values.

How do you reason with different function types in order to make sense of contexts such as the one presented in the task in figure 4.1 (page 75)? You are provided with four functions with different algebraic structures. How do you analyze functions that possess different structures? What information or strategies would be useful in examining these different function types? Often when students are given different functions, and in particular different types of functions, a typical strategy involves inputting different values for each function and examining the results (see table 4.1).

Table 4.1: Initial Selected Values for Function Comparison

x	f(x) = 2x	g(x) = x + 2	h(x) = x²	l(x) = 2ˣ
0	0	2	0	1
1	2	3	1	2
2	4	4	4	4
3	6	5	9	8
4	8	6	16	16

What conclusions can you draw from the values listed in table 4.1? What do you notice about the values chosen for x? How do you make the connections between the values of x, the function values, and their meaning for the population growth over time? The outputs from the four given functions are all equal in the second row of table values—namely when x equals 2, which happens to be a commonly selected value to test in order to determine which function is greater. Would the values in table 4.1 represent sufficient information to determine which function describes the greater population growth? Notice how the answer to that question depends on the values of x that you chose to examine. You need to think about what knowledge and strategies are needed to recognize whether the values in table 4.1 represent sufficient information to determine which function has the greatest growth in function values.

Note how some of the values for x negate the idea that the linear function with the greater slope ($f(x) = 2x$) consistently produces greater function values than the linear function with the lesser slope

($g(x) = x + 2$). A quick look at the graph of these two linear functions may also provide insight in distinguishing between function values and slope (see figure 4.2).

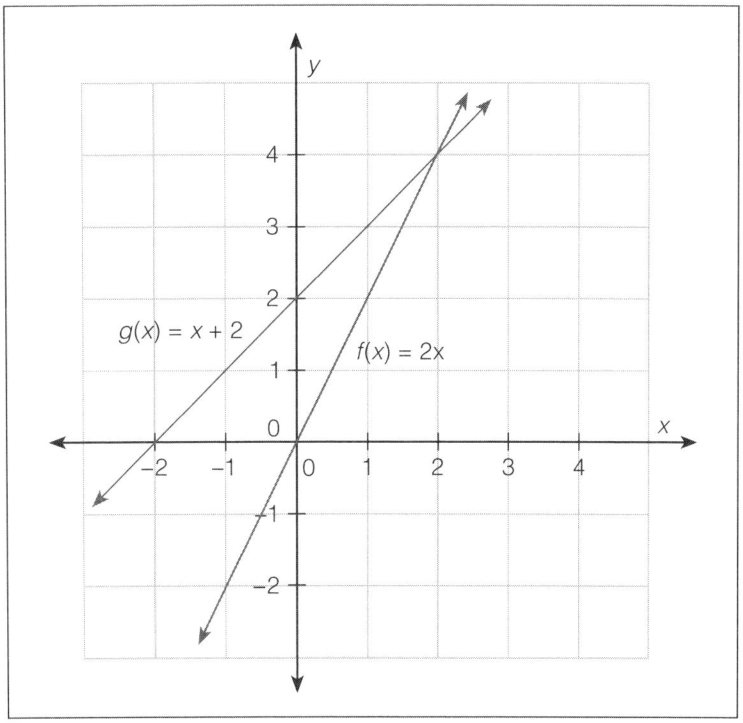

Figure 4.2: Graph of the linear functions from the population growth through function types task.

How can you deepen your exploration of how the functions compare based on the fact that the functions take on equal output values at $x = 2$? You might conclude that table 4.1 does not provide sufficient information to answer the question fully and accurately. Consider table 4.2.

Table 4.2: Further Exploration of Values for Function Comparison

x	$f(x) = 2x$	$g(x) = x + 2$	$h(x) = x^2$	$l(x) = 2^x$
−2	−4	0	4	0.25
−1	−2	1	1	0.5
0	0	2	0	1
1	2	3	1	2
1.5	3	3.5	2.25	≈2.83
2	4	4	4	4
3	6	5	9	8
4	8	6	16	16

How does table 4.2 provide a different perspective compared to table 4.1? Why do you suppose the value of 1.5 was chosen for x?

You may have noticed that in the first table, it appears that one linear function ($g(x) = x + 2$) takes on the same value as the quadratic function ($h(x) = x^2$) twice, whereas the other linear function ($f(x) = 2x$) only shares one value with the quadratic function. An examination of the second table of values and the graph of these functions (see figure 4.3) highlights that there are actually two points of intersection in each case, demonstrating the importance of examining multiple representations to compare functions.

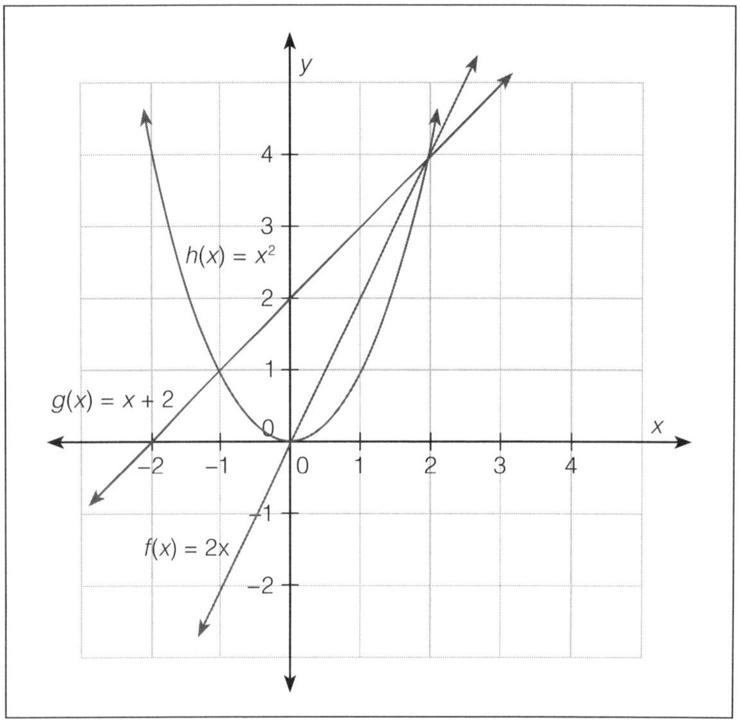

Figure 4.3: Graph of the linear and quadratic functions from the population growth through function types task.

Notice how these representations work together to provide insights into the different functions. You also have the connection to negative x-values representing years before 2010. Students need to understand such connections in order to recognize, use, and correctly apply each type of function in mathematical and real-world contexts.

How does the growth rate for the exponential function $l(x) = 2^x$ compare to the growth rates of the linear and quadratic functions? How is it possible that the exponential and quadratic function types take on the same value on two separate occasions? Consider how the function values change as the x-values increase by one-unit increments. For one linear function, $f(x) = 2x$, as the x-values increase by one unit, the function (y-values) consistently increases by two units. In the given context, that would mean that for this linear function, the population changes by 2,000 people per year since the function values are measured in thousands of people. In the other linear function, $g(x) = x + 2$, as the x-values increase by one unit, the function (y-values) again consistently increases, this time by one unit, or 1,000 people per year. In both cases, the population growth varies by the same amount with each passing year. Compare this behavior with the quadratic function $h(x) = x^2$ and the exponential function $l(x) = 2^x$. What type of change in the function (y-values) do you notice per unit increase in the x-values? For the quadratic function, the transition of the rate of change from negative to positive is evident in table 4.2 (page 77). The rate of change between two values of x depends on the particular x-values that are selected. For the exponential function, the rate of change is consistently increasing, and the rate of change grows by a constant multiplier of 2 for each whole unit increase of x. Therefore, the question of which city has the greater population growth is answered by the fact that the growth rate relationship varies among the different cities and that the greater growth depends on when the growth rate is measured.

All of these comparisons were initially made using a set of values that were organized in a table format, with occasional support from graphs. What are the benefits of having these numerical values represented in a table? What are the disadvantages or limitations? How did this analysis compare to the insights gained from interpreting the graphical representations? This task and the context of population change for the four different cities provide an entry point into how to use both mathematical tools and context to examine, analyze, and reason with different function types. This is a major focus of high school mathematics. What tools and representations enable you to build understanding of different mathematical functions?

These questions will be addressed as you continue to consider the significance of exploring concepts before procedures and the importance of having multiple approaches in order to reason with mathematics. In the next chapter, you transition from exploring tasks in context to modeling tasks using your knowledge of mathematics. Throughout this book, the perspective is consistent that concepts should be discussed before—but not in place of—procedures. Once a conceptual understanding has been emphasized, procedures and strategies are developed that support mathematical thinking.

The Progression

Similar to geometry, mathematical concepts in algebra are built on earlier experiences. Analyzing function types connects to earlier high school coursework. Students first learn about functions and function types using specific numbers and values in table format and examine graphical connections for each function type, as you read in chapters 1 and 2. As the following progression shows, developing understanding of functions builds on and extends previous knowledge of functions by analyzing additional functions using different representations.

- Analyze functions of different types using multiple representations.
- Create functions that model a relationship between two functions (for example, rational functions).
- Build new functions from existing functions (for example, inverse functions).
- Combine and compose functions to create new functions.

High School

Students see and analyze characteristics of linear, quadratic, exponential, logarithmic, rational, and trigonometric functions. A major part of the progression entails the analysis of different function types and the construction of new functions from existing functions. For instance, students explore scenarios in which two linear functions are multiplied together. Eventually, they investigate contexts in which two linear functions are divided to create rational functions, which is an extension of students' study of proportional reasoning from the middle grades.

By understanding multiple representations of functions, students begin to appeal to different aspects of functions to deepen their understanding. They use the ideas of the algebraic structure of functions to relate quadratic functions with square root functions, exponential functions with logarithmic functions, and rational functions with functions involving polynomials. Similarly, students create new functions from existing functions by extending notions involving x^2 or y^2 (such as quadratic functions creating parabolas) to graphs that involve both x^2 and y^2 simultaneously (such as circles, ellipses, and other conic sections).

The Mathematics

A focus on function types must include connections to parent functions. When students make sense of parent functions and transformations of those functions, students are able to make inferences regarding the behaviors of functions and link characteristics of different types of functions together. Knowing the general behavior of a function helps in determining if the application of that function to model a real-world context is a reasonable choice. High school students must be able to determine the impact of parameters on function behavior, make connections through representations, connect types of functions using symmetry, and explore rational functions.

Determining the Impact of Parameters on Function Behavior

Functions are often presented, explored, and analyzed as families. Families of functions involve grouping functions by specific parameters to provide a deeper understanding of how variations in these parameters impact the behavior of the graph of a function. For instance, in early high school coursework, students examined $y = mx + b$ and discussed the role of parameters m and b on the behavior of the line. The learning progression extends this analysis to various function types including quadratic (see figure 4.4), exponential, and logarithmic functions.

DO NOW

Graph the following functions:

$f(x) = (x - 3)^2 - 2$ $g(x) = (x - 3)^2 + 2$

$h(x) = 2(x + 3)^2 + 2$ $j(x) = 2(x + 3)^2 - 2$

How do the changes in the values of the parameters of these functions affect their graphs?

Figure 4.4: Transformations of quadratic functions task.

Students explore the impact of changing different parameters of the quadratic family of functions presented in vertex form $f(x) = a(x - h)^2 + k$. The impact of changing the numerical value of the parameter a affects how narrow or wide the graph will appear. In fact, the parameter a provides a sense of how quickly the quadratic function increases or decreases, identifying the rate of change of the quadratic function. Additionally, understanding the role of h and k as they relate to the vertex of the parabola will link to the study of axis of symmetry and locus of points, and extend to the study of other conic sections.

In the case of $f(x) = (x - 3)^2 - 2$, the parent graph of $y = x^2$ is shifted three units to the right and two units down. This provides the reasoning for why the vertex shifts from (0, 0) to (3, –2).

In this example, how do you justify that in order to get the x-coordinate of the vertex, you use a positive 3 when the expression is $x - 3$, while you take the value of 2 without altering the sign in order to find the y-coordinate. Recall the discussion from chapter 2 about the role of $(x - h)$ and k. If you set the expression $(x - 3)^2$ equal to 0 and solve for x, what value would you get for x? This provides the mathematical reasoning to determine the impact of the parameter h. Next, examine the role of the parameter k. Notice that in the function $f(x) = (x - 3)^2 - 2$, the impact of the –2 is to decrease all of the y-values by two units. The other given functions in this task are vertically translated based on the value and the sign of k. This process of reasoning allows you to understand the role of the parameters h and k both numerically and graphically.

Another exploration that you can investigate with these parameters is how to create a graph of a quadratic function that opens downward instead of upward. Through the use of multiple representations, students could explore ways in which the parameter a impacts the behavior of quadratic functions. Looking at the parabolic shape of the graph of a quadratic function, students again analyze the role of the sign of the parameter a to identify that a negative value of a creates a parabola that opens downward, linking this idea to the vertex form of the equation.

How could you use your knowledge of quadratic functions and the associated parameters to create a parabola that opens downward with the same vertex as $j(x) = 2(x + 3)^2 - 2$? You can see from the earlier discussion that the vertex of this parabola should be (–3, –2). While the coefficient value of 2 indicates how wide the graph appears and how quickly the y-values increase, the sign of the parameter a indicates whether the parabola opens up or down. As such, if you were interested in having the parabola with the same vertex and width and opening downward, you would want the leading coefficient to be negative. By making $a = -2$, you would have $k(x) = -2(x + 3)^2 - 2$, which would yield the desired result (see figure 4.5).

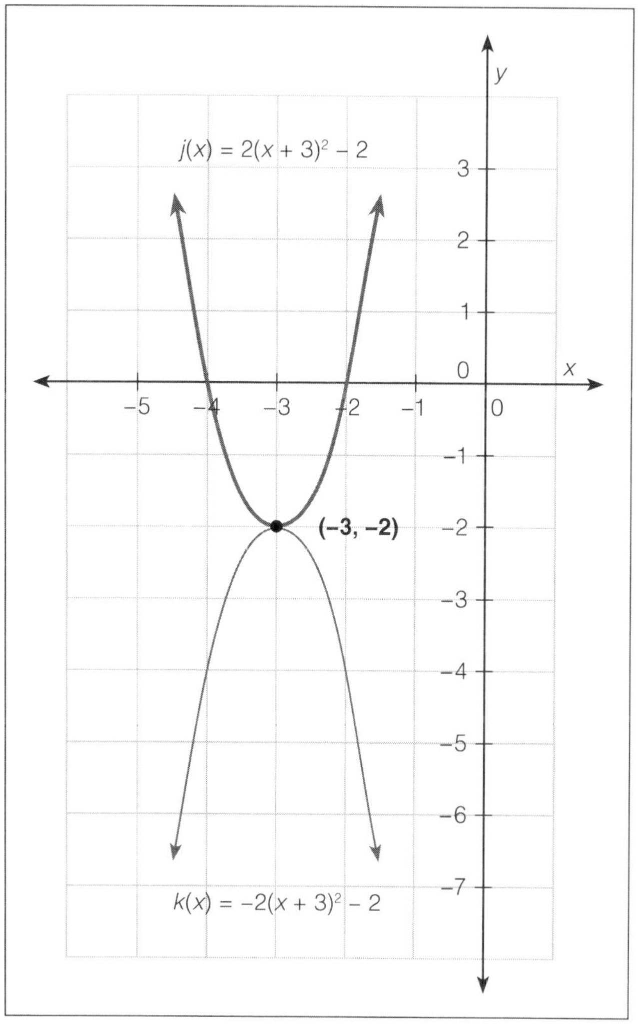

Figure 4.5: Quadratic functions in vertex form with leading coefficient $a = 2$ and $a = -2$.

Consider the parameters in the family of exponential functions, $f(x) = ab^{(x-h)} + k$, where b cannot be 0. Why do you see the same parameters in this representation as you did in the quadratic function? While there are other ways to explore the role of parameters involving the family of exponential functions, this form connects to the transformations you investigated using quadratic functions. In the quadratic function family, the coefficient a impacts the curvature as well as whether the parabola opens upward or downward. In addition, the role of the parameters h and k identify horizontal and vertical shifts, respectively. For this family of exponential functions, the parent function is $f(x) = b^x$. In the form that includes the parameters, as h changes, the graph shifts h units horizontally. Similarly, as the parameter k changes, the graph shifts k units vertically. See the example of these shifts with a family of exponential functions based on $y = e^x$ (figure 4.6).

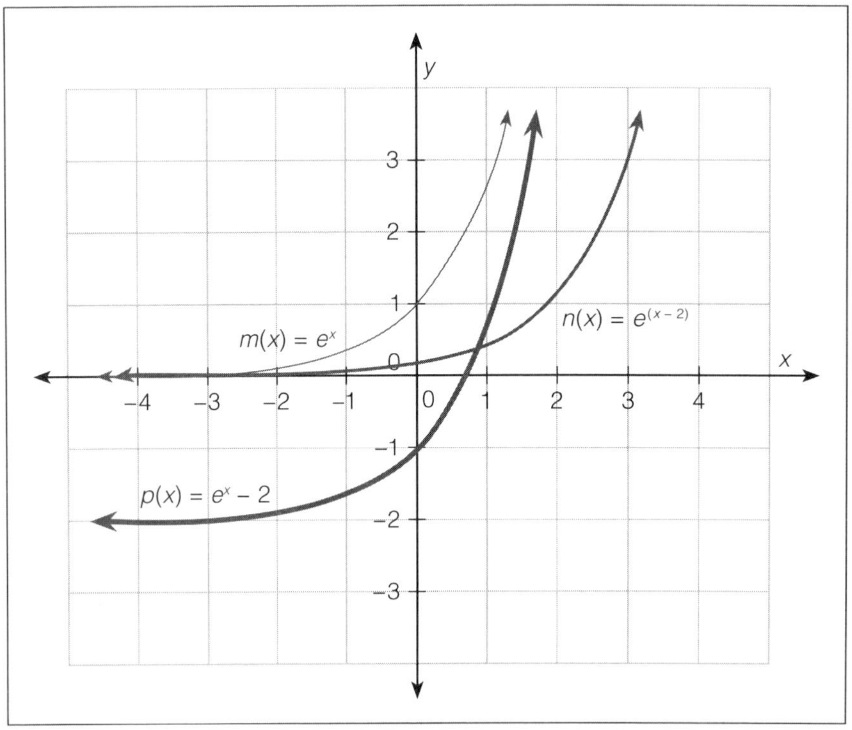

Figure 4.6: Impact of parameters on $y = e^x$.

From the graph in figure 4.6, you see that the function $n(x) = e^{(x-2)}$ is shifted two units to the right of $m(x) = e^x$, and $p(x) = e^x - 2$ is shifted two units down.

The pattern of changing these parameters is maintained between quadratic and exponential functions and supports students' study of various function types throughout their experiences in high school mathematics. This development is powerful as it demonstrates the connections across topics in mathematics using sense making and reasoning rather than strictly relying on procedural methods and memorization. How will the parameters in the family of logarithmic functions—$g(x) = a\log_b(x - h) + k$—impact the parent function of $f(x) = \log_b x$? Again, the h and k shift the parent function horizontally and vertically, respectively, and the parameter a transforms the function by stretching or compressing the parent function, or by reflecting it across a horizontal line.

Similar to quadratic, exponential, and logarithmic functions, you can also examine the role of parameters in trigonometric functions. Explore the task in figure 4.7.

DO NOW

Predict the graph of this sine function:

$f(x) = 2\sin(x - \frac{\pi}{2}) + 1$

What is the range of values for this function?

Figure 4.7: Predicting the behavior of a sine function.

What roles do the parameters *a*, *b*, *c*, and *d* play in the trigonometric function $y = a\sin(bx - c) + d$? Why do you think there are no parameters *h* and *k* in this family of functions? In the case when parameters *a* and *b* are equal to 1 and *c* and *d* are 0, you have the standard or parent sine function, $f(x) = \sin(x)$. The sine function follows a cyclical pattern, where function outputs—the range—follow a pattern between –1 and 1. The length along the independent variable axis of one complete repetition of this pattern is called the *period.* For the parent sine function, the period is 2π (see figure 4.8).

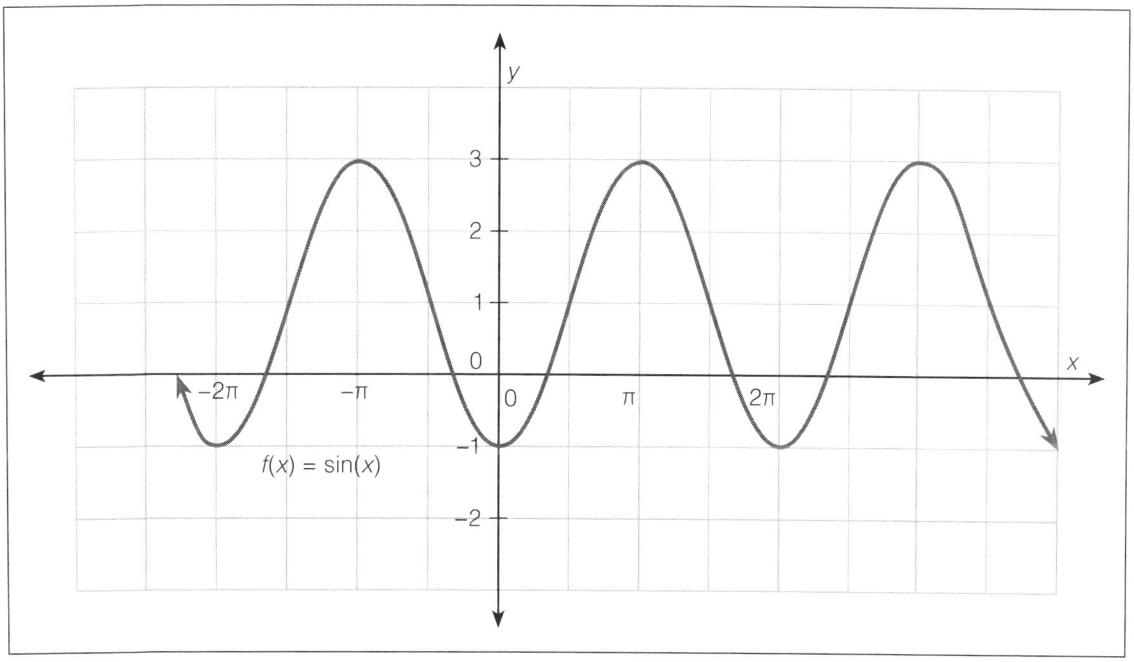

Figure 4.8: Sine function in standard position with period of 2π.

How does the parameter *a* affect the behavior of the function? The parameter *a* impacts the amplitude of the function—affecting the range of the function—as well as whether the graph is in the standard position (see the graph in figure 4.8) or is reflected across the baseline (which would be the *x*-axis in figure 4.8). This role of parameter *a* is consistent with the role of *a* in the quadratic and exponential functions studied previously, as it impacts how fast the values of the function change. The values of *c* and *d* in $y = a\sin(bx - c) + d$ determine the horizontal and vertical shifts of the function, similar to the role of *h* and *k* in the functions studied earlier. However, as the sine function does not have a vertex, the parameters are identified differently, though their role is similar. For the sine function, the parameter *b* impacts the period of the function. When *b* is greater than one, the output of the sine function will repeat faster and the cyclical pattern is condensed; when *b* is less than one, the output of the sine function will repeat more slowly, and the period is expanded. Now, compare the parent sine function with $f(x) = 2\sin(x - \frac{\pi}{2}) + 1$. The parameter *a* takes on the value of 2, which indicates the amplitude of the function is twice as large as before. What can you say about the parameters c and d compared to the parent function? Here, the parameter c is $\pi/2$, which means that the function is shifted $\pi/2$ to the right from its standard position. The value of d here is +1, indicating that the entire graph is vertically translated in the positive direction by one unit. Considering the effect of the vertical translation (+1) and the amplitude (two units), this trigonometric function now ranges between the y-values –1 and 3 and, since the value of b is 1, maintains the normal period of 2π. Including the shift of $\pi/2$, the graph of $f(x) = 2\sin(x - \pi/2) + 1$ is shown in figure 4.9.

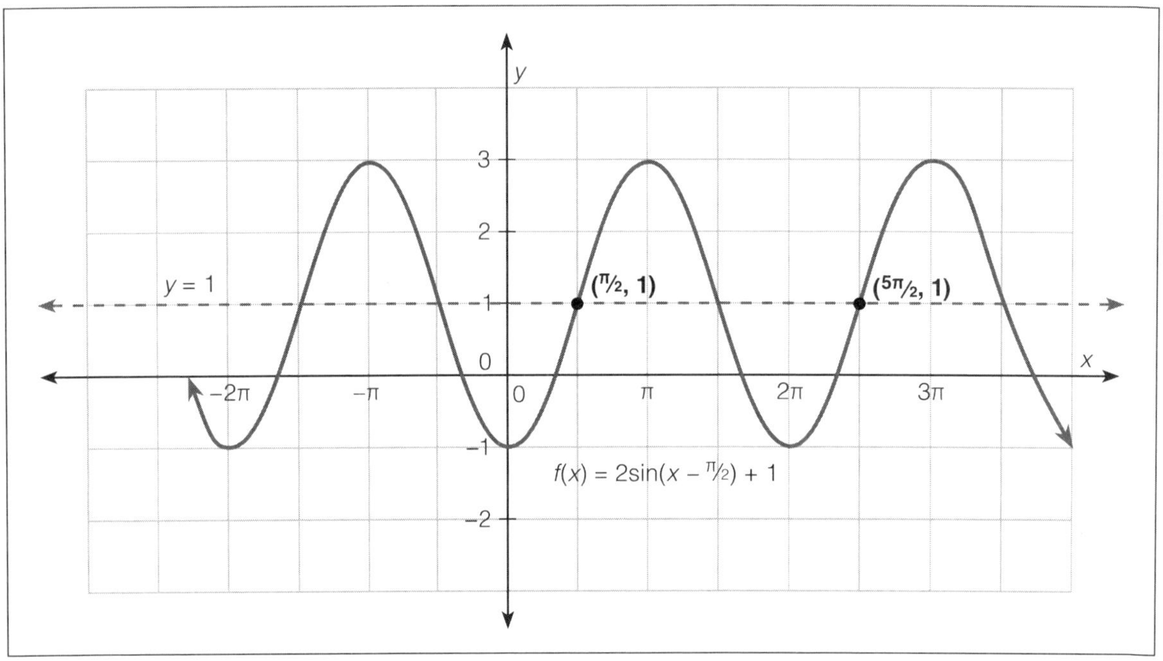

Figure 4.9: Graph of $f(x) = 2\sin(x - \pi/2) + 1$.

Making Connections Through Representations: Graphic and Algebraic

Throughout the algebra sequence in high school, conceptual connections can be made through an emphasis on multiple representations, building on what students have learned in the first year of their algebra study. Think about the connections across the algebra concepts studied in high school in the task in figure 4.10.

When you graphically explore the product of $f(x) = x + 1$ and $g(x) = x + 2$, what connections do you observe? The relationship between the points in the linear examples generates the range of the product of these functions and creates a range that is different from the range of the initial functions. If you think about the values of each linear function, you realize that the product of the two functions will only be negative at x-values when exactly one of the values of the two linear functions is negative. This means that the product of these two functions will only be negative between the x-values of –1 and –2 (see figure 4.11, page 86).

In addition, conceptual connections can be made as the x-intercepts of the linear functions $f(x) = x + 1$ and $g(x) = x + 2$ are at $x = -1$ and $x = -2$, respectively. The x-intercepts of the quadratic function remain at the exact same points because the quadratic function is the product of the two earlier linear functions.

DO NOW

Examine the graphs of $f(x) = x + 1$ and $g(x) = x + 2$ provided below.

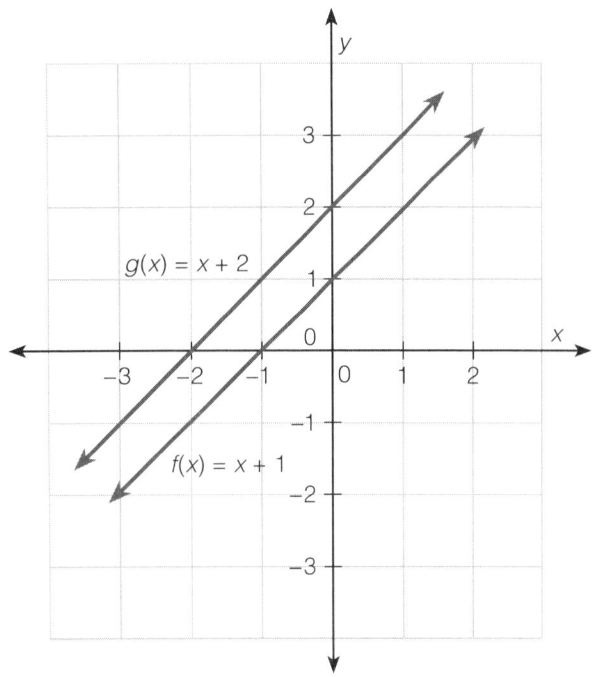

1. Predict when the graph of the product of these two functions will be below the x-axis.
2. Graph the outcome of multiplying these two functions.
3. Determine the connections between the x-intercepts in the graphs.

Figure 4.10: Graphical connections of the product of linear functions.

A value of zero for each linear function will be a zero for any function in which that linear function is a factor. When new functions are built based on existing functions, previous understandings are extended and applied in new settings. From the graph, you see that the vertex of the parabola is directly between the two x-intercepts. Why is this so? The two x-intercepts for this parabola, which occur at $x = -1$ and $x = -2$, are symmetric about the vertex. As these points indicate symmetry, the x-coordinate of the vertex will be halfway between the x-intercepts, at $x = -1.5$. Evaluating the two linear functions at this x-value and finding their product shows that the vertex is $(-1.5, -0.25)$ (see figure 4.11, page 86).

Connecting Types of Functions Using Symmetry

Throughout the middle grades and into high school, students learn about the notion of symmetry as it pertains to patterns, shapes, and geometric properties. This emphasis is maintained and extended as students use symmetry to develop ways to connect functions. How can students make sense of the relationship between quadratic and square root functions? Students have evaluated the square root of x primarily using numerical approaches and tables of values. Compare this approach to the task in figure 4.12 (page 86).

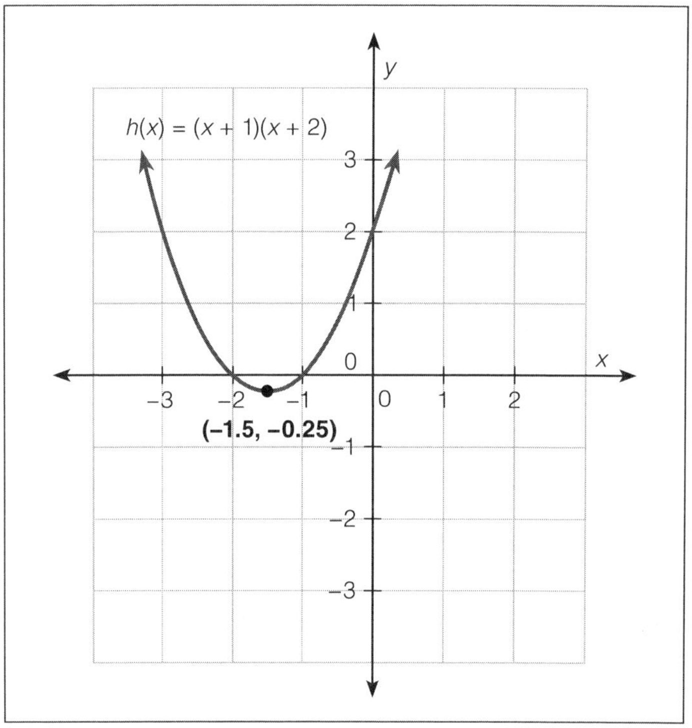

Figure 4.11: Quadratic function explored as the product of two linear functions.

DO NOW

Consider the graph of $f(x) = x^2$ with the restricted domain $[0, \infty]$.

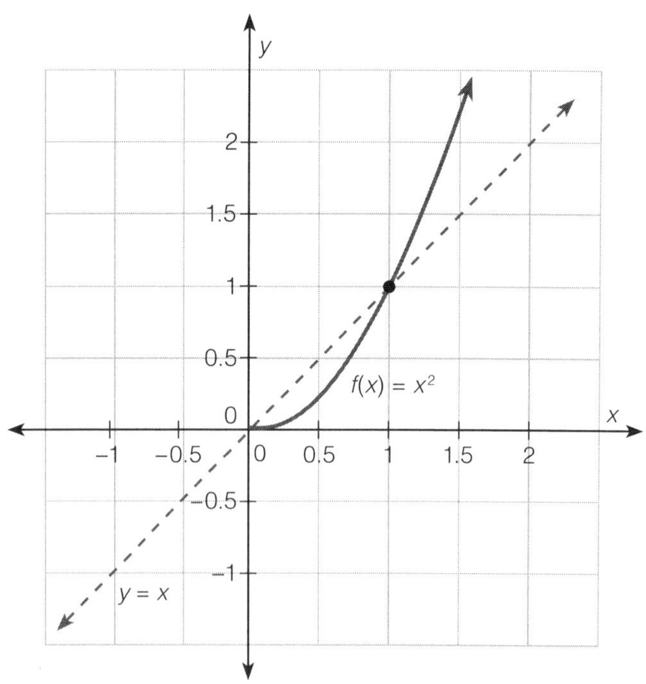

Predict the function resulting from reflecting $f(x) = x^2$ about the line $y = x$.

Figure 4.12: Exploring the relationship between functions through reflection.

What relationship do you notice numerically, graphically, and algebraically when you reflect all of the points from $f(x) = x^2$ across the line $y = x$? Why does this relationship exist? Realize that the line $y = x$ is the collection of points equidistant from both the x- and y-axes. In reflecting across the line $y = x$, you are transforming points from being a certain distance from the origin relative to the x-axis to be that same distance from the origin but now relative to the y-axis. You are transforming the coordinates of all points of a given function (x, y) to correspond to the points of the inverse of the function (y, x). Consider the graph of the pre-image of $f(x) = x^2$ and reflected image (see figure 4.13).

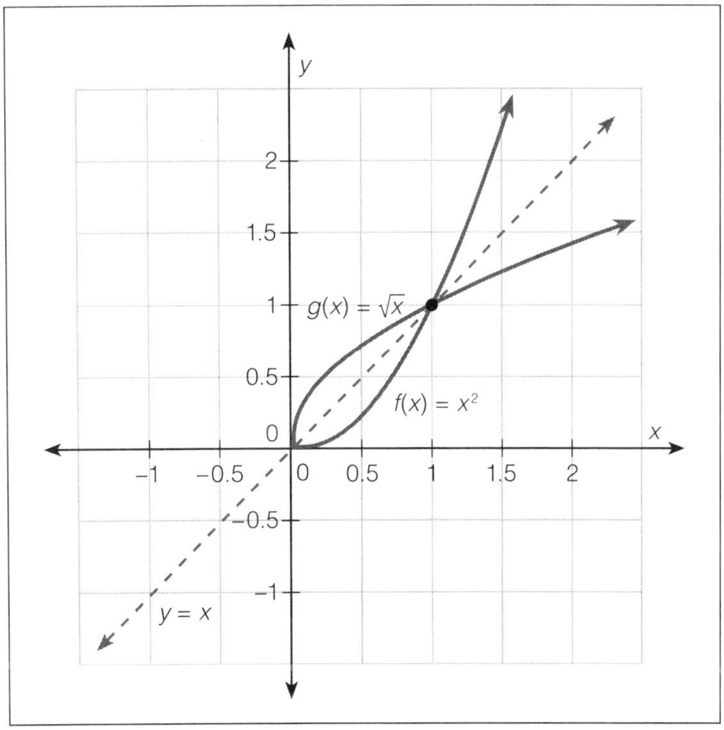

Figure 4.13: Symmetry of a quadratic and a square root function.

Squaring a number (and because of the restricted domain, a positive number) and then taking the square root returns the value of the original number, one "undoes" the other. In mathematics, functions that operate in this manner are inverses to one another. Recall the discussion of inverses from chapter 1 (page 28). Composing these two functions—placing one function within the other—yields $(f \circ g)(x) = f(g(x)) = f(\sqrt{x}) = (\sqrt{x})^2 = x$. In other words, applying one function, $g(x)$, directly followed by the next function, $f(x)$, returns the original input value, x. Conversely $(g \circ f)(x) = g(f(x)) = g(x^2) = \sqrt{x^2} = x$. Therefore, even if you were to compose these two functions in the reverse order, you would still undo the impact and get the original value of x. This idea of symmetry—both about the line $y = x$ and in the composition of functions—assists in thinking about inverses beyond just exploring the connection between quadratic and square root functions.

How could this work in context? You can engage in attaching mathematical meaning to the word *inverse* where you invert the role and location of x and y. Imagine having the revenue function $R(x) = -20p^2 + 1000p$, where $R(x)$ is the revenue generated at a particular product price p. With the given function, you are able to determine the revenue from a given unit price. You may also want to determine a particular product price that would generate revenue. In this example, you might want to determine a product price to generate revenue of \$20,000 or \$40,000. You could switch the role of the revenue and price variables to determine the appropriate product price. The function that relates the independent variable of product price to revenue would be the inverse relationship of the revenue function.

Consider the function $f(x) = x^2$. The point (2, 4) satisfies the function $f(x)$ and is on the graph of this function. The inverse function, $g(x) = \sqrt{x}$, should contain the same point with the order inverted. Hence, the point (4, 2) satisfies the function g and is indeed on the graph of $g(x) = \sqrt{x}$ (see figure 4.14).

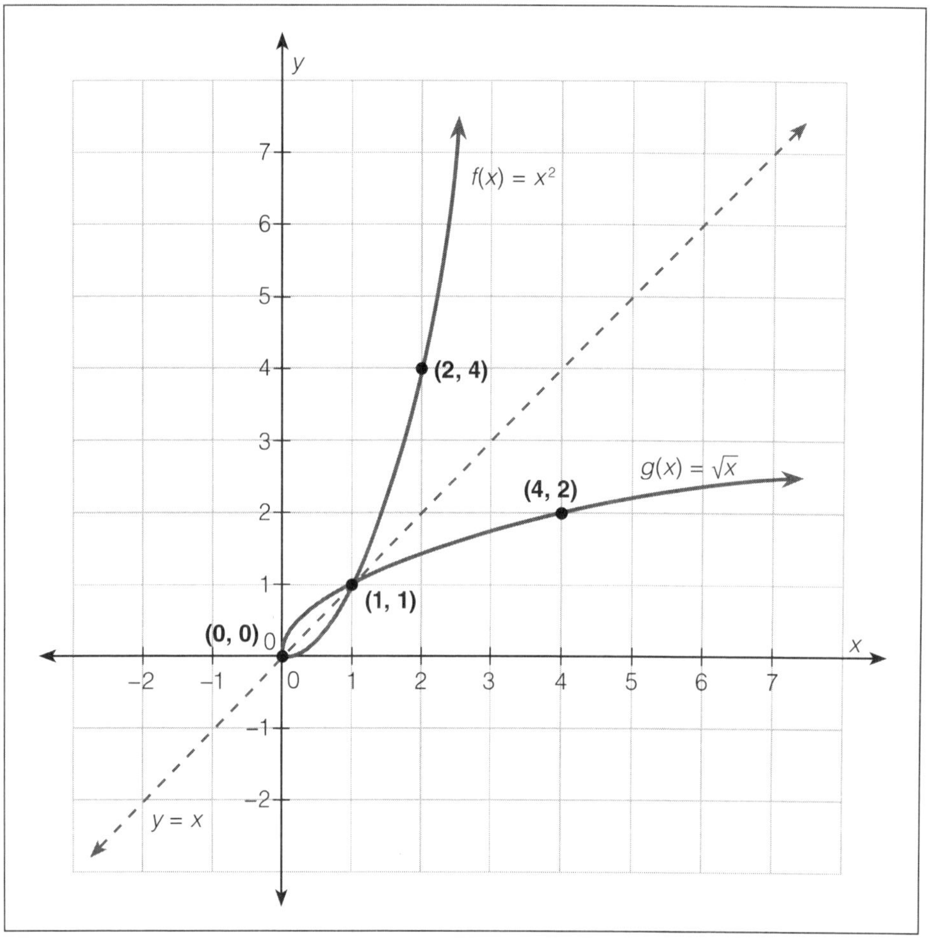

Figure 4.14: Graphical connections of a quadratic and a square root function.

Notice that on some occasions, the use of convenient points such as (0, 0) and (1, 1) is actually not helpful in demonstrating and connecting inverse functions, as these points are common to both functions. This relates back to the discussion regarding the task provided in figure 4.1 (page 75) about the appropriate selection of values to analyze functions. While (0, 0) and (1, 1) do indeed satisfy both functions $f(x) = x^2$ and $g(x) = \sqrt{x}$, by selecting these points you would miss the opportunity for students to explore the inversion of the x- and y-values. The selection of a point such as (4, 2) provides an opportunity for a richer understanding of function inverses, similar to the relationship between revenue and price.

By connecting the numerical and graphical aspects of the quadratic and square root function, you have demonstrated how mathematical modeling can be used to analyze mathematical events. Why did it make sense to explore only non-negative inputs for $f(x) = x^2$? What would happen if you examined an unrestricted domain of the function? As you may have noticed, you need to restrict the domain of this function in order for a one-to-one functional relationship to exist between the function and its inverse. Limiting the domain of this function is necessary in order for the inverse to also remain a function. The fact that some values of the range for $f(x) = x^2$, such as 4, have two domain elements, –2 and 2, does not preclude the relation from being a function. However, in examining the inverse function, switching the sets for domain and range causes a problem. When the value of 4 in the domain links to two values in the range, –2 and 2, then the relation is not a function. This is why there are times when you need to limit the domain of a function in order to have a one-to-one function (see discussion of one-to-one functions in chapter 1). Consider the situation graphed in figure 1.16 (page 29). In this context, the inverse function determines the amount of time it takes Javier to travel a particular distance. Javier is traveling at a constant speed. As the function is one to one, you can also determine the travel time for a specific distance. Other important real-world examples of one-to-one functions are social security and credit card numbers. Consider the issues that would arise if social security numbers were not uniquely assigned or if credit card numbers were assigned to multiple people. Moving forward, it is important to consider how to use information about one function type to create and modify other types of functions.

Students also extend their analysis of different types of functions to include aspects of exponential and logarithmic functions. The exploration of these functions should include numerical, graphical, and functional experiences in order to deepen students' understanding. Explore the relationships between exponential and logarithmic functions (see figure 4.15, page 90).

How is this task similar to the connection between the quadratic and square root functions? Consistent with the previous discussion, the graphical symmetry between the exponential function $f(x) = e^x$ and the natural log function $g(x) = ln(x)$ demonstrates an inverse relationship between these functions. Observe the relationship between the point $(1, e)$ in the exponential function and the point $(e, 1)$ in the natural logarithmic function. How do the points and graph of each function relate to one another? How does the line $y = x$ support your understanding of inverse functions? Connecting the relationship between domain and range for functions and their inverses allows you to delve into functional and graphical interconnections by extending the notions of symmetry and reflection. With the gradual enhancement of mathematical ideas, students can reason with functions and make sense of the manner in which the coordinates of the points are inverted and the graphs are symmetric across $y = x$, and the ways in which exponential functions with base e and natural logarithmic functions, which also have the base e, are connected. In this manner, functional understanding can be developed systematically and purposefully while exhibiting the structure and regularity that is at the heart of mathematics. Here, students should be engaging in Mathematical Practice 7, "Look for and make use of structure," and Mathematical Practice 8, "Look for and express regularity in repeated reasoning."

How can you express a nonfunctional relationship using functions? Consider the representing nonfunctions with functions task in figure 4.16 (page 91).

DO NOW

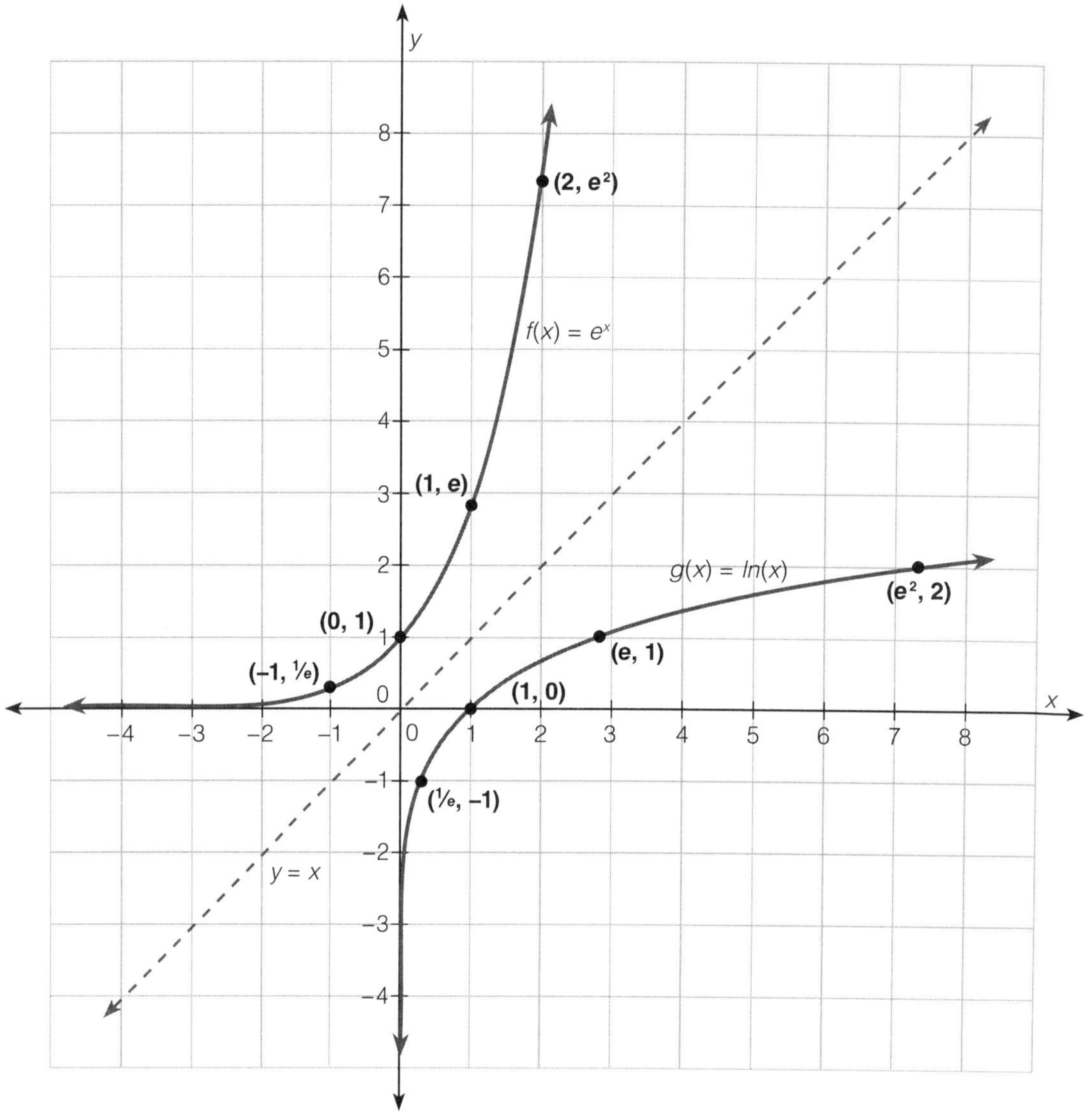

Describe the relationship between these two functions.

Figure 4.15: Connecting exponential and natural logarithmic functions.

At first glance, you might be tempted to respond that the graph does not represent a function. You would be correct in that response; it *can*, however, be described by more than one function. A common first approach would be to break up the problem into two graphs where a known function type could represent each piece. You may recognize the graph in the task and break it apart into two separate graphs that have the point (2, 0) in common.

Knowing the parent function for $f(x) = \sqrt{x}$ enables you to analyze the shape and realize that it is the parent square root function translated horizontally. By recognizing a shift of two units to the right, you can adjust the equation to be $f(x) = \sqrt{x - 2}$ in order to attain the top half of the desired graph. This is where your experiences working with the parameters of different functions provide a strategy to use for

DO NOW

How can you describe the graph below using functions?

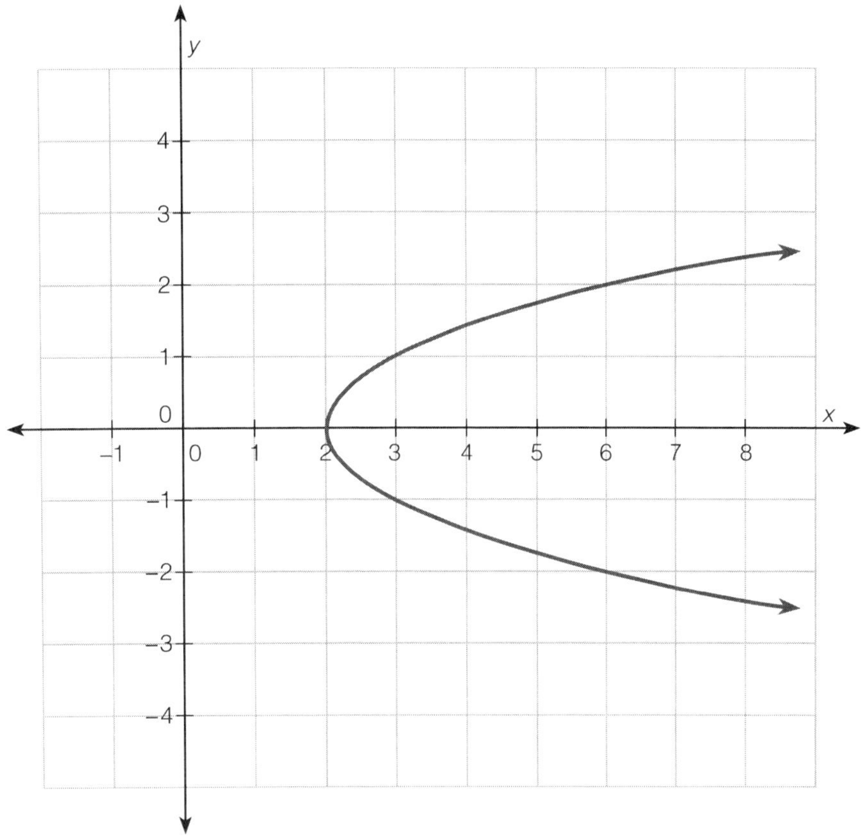

Figure 4.16: Representing nonfunctions with functions task.

this task. Similarly the bottom half of the graph involves a reflection of your selected function about the x-axis, as the y-values are now negative. This function is represented by $g(x) = -\sqrt{x-2}$. Using symmetry and translations of functions enables you to decompose the original graph into two pieces that can be represented as functions (see figure 4.17, page 92). This ability to break down problems using previous knowledge and experiences is the type of analysis and sense making that you need to develop in your students.

Another approach to arriving at the function that models the graph in figure 4.16 involves relying on a thorough understanding of parabolas. Earlier in this chapter (see figure 4.11, page 86) you explored the nature of quadratic expressions and the symmetric relationship demonstrated in the graph. You may notice that the graph in figure 4.16 displays a symmetric relationship about the x-axis, with a vertex at the point (2, 0). This presents an opportunity to be flexible in your thinking about functions. Typically, you view relationships in the form y as a function of x. In this example, the points (4, 2) and (4, –2) could indicate the y-values as independent variable values to obtain the resulting x-values. That is to say, given $y = -2$, the dependent variable is $x = 4$, and similarly, given $y = 2$, the output also provides an x value of 4. The mathematics here suggests that one way to think of this relation is that x is now a function of y. The function involves given particular y-values as independent variable values, you need to square those

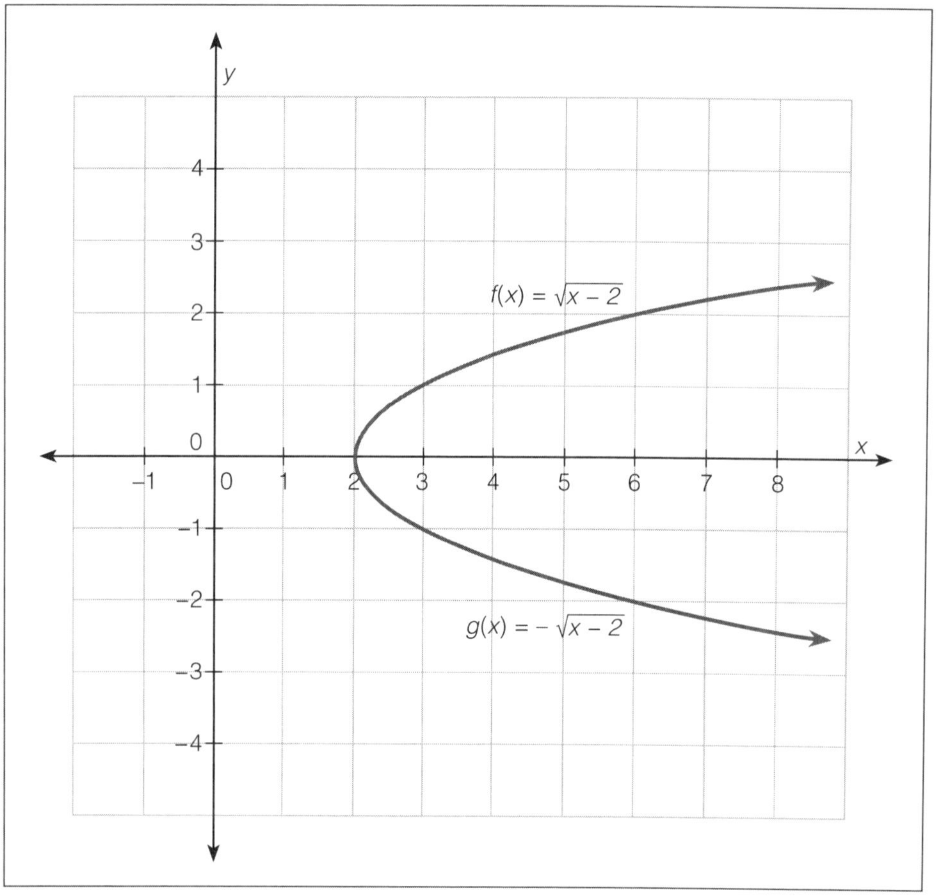

Figure 4.17: Two functions to represent a nonfunction.

values to get the dependent variable value represented by the variable x. Hence, the parent function representation would be $x = y^2$ (if you were allowed to switch the independent and dependent axes). A quick examination, however, tells you that the graph demonstrates a shift of two units, and as such, you can reason that your equation would be $x = y^2 + 2$. Notice how the translation in this case occurs in the location of k, as you have broadened your understanding of a functional relationship in thinking of the relationship as x as a function of y. You can test your conjecture and become more certain of $x = y^2 + 2$ as the algebraic representation of the graph provided in the task.

Lastly, consider the graph in figure 4.16 (page 91) from a numerical standpoint. Some students may find it easier to approach the task in this way (see figure 4.18).

You see the table on the left explores the relationship starting with different x-values. The table on the right instead begins with different y-values. Consider how the mathematical meaning for functions allows you to examine relations and functions through multiple perspectives. The values on the left table represent a set of points that are mapped to different y-values. However, the x-value of 3 is paired up with both y-values of 1 and −1. As there is more than one output for an individual input of x, this representation on the left is not a function (also see chapter 1). This realization could be an indication for students that the graph in the task in figure 4.16 is not a function of x.

x	y
2	0
3	1
3	−1
6	2
6	−2

y	x
−2	6
−1	3
0	2
1	3
2	6

Figure 4.18: Two tables representing ordered pairs from the task in figure 4.16.

Additionally, many students would immediately dismiss the graph by saying it is not a function since the graph does not pass the vertical line test. This mathematical connection allows for you to highlight Mathematical Practice 6, "Attend to precision." It is true that the graph does not pass the vertical line test; however, the vertical line test specifically determines if a graph represents a function *of* x. You need to seize this opportunity to push for the precision that is needed and arrive at a deeper understanding of when a relation represents y as a function of x, or whether it is possible for the relation to represent x as a function of y. Note that the graph in figure 4.16 can be represented by the equation $x = y^2 + 2$, which, in fact, represents x as a function of y. This type of reasoning becomes important as students make sense of graphs of conic sections and other coordinate systems in precalculus and calculus.

Examine the table of values on the right side of figure 4.18. Note how the x-values are symmetric about the ordered pair (0, 2). The y-values of 1 and −1 both have the same x-value as do the y-values of −2 and 2. The table of values reinforces the connection to thinking of x as a function of y. This discussion of connecting different representations to reinforce this flexible view of functional relationships relies on an emphasis on Mathematical Practice 7, "Look for and make use of structure." You based the reasoning in this problem on the structure of quadratic functions, symmetry, and how multiple representations can aid in looking for and making use of such structure. The fact that the points are symmetric about the point where the y-value is 0 replicates what students have previously experienced with quadratic functions that open upward and downward. Here, you appeal to this structure to extend this understanding in a new task while continuing to reason with different types of functions (quadratic, square root, and so on). While the nature of this task involves persistence as outlined in Mathematical Practice 1, "Make sense of problems and persevere in solving them," the goal is to address the content conceptually prior to procedurally so students can explore multiple ways to think through and use the structure of quadratic functions while attending to precision.

Exploring Rational Functions

Another type of function that is discussed in high school mathematics involves rational functions such as $y = {}^{(x)}/_{h(x)}$, where $f(x)$ and $g(x)$ could represent linear or other polynomial functions. Consider the task in figure 4.19 (page 94).

DO NOW

Given the function $f(x) = \frac{2x - 5}{x + 1}$, consider the domain and key aspects of the graph of this function. Sketch a graph for this function.

Figure 4.19: Graphing rational functions task.

What do you know about the rational function provided? What is the domain? What are the *x*- and *y*-intercepts for this graph? These are important characteristics of graphs of rational functions that can be explored early in the investigative process, connecting to what is already known from working with other functions. What happens to the function at values of the domain where the function does not exist, such as values that would lead to dividing by 0? For the given function, the value of $x = -1$ causes the graph to be discontinuous, leading to an asymptote. Are there vertical or horizontal asymptotes for this function? How would you determine them? This is an opportunity to explore the role of asymptotes in rational functions before formally defining this term in future mathematics courses.

You can see that the domain should exclude the point $x = -1$ because when you divide by 0, the value of the function will be an indeterminate value. As the *x*-values approach –1 from the left, the denominator is negative but gets closer and closer to zero. Since the numerator is also negative, the function value gets larger and larger, growing toward positive infinity. When the *x*-values approach –1 from the right, the denominator is positive but gets closer and closer to zero. Since the numerator is also negative in this case, the function value gets smaller and smaller, moving toward negative infinity. As such, graphically, the equation $x = -1$ would represent a vertical asymptote for this function, a place where the graph of the function is not connected. A vertical asymptote occurs when the function value approaches positive or negative infinity as the *x*-values get infinitely close to a value from one side. In this case, as the *x*-values get infinitely close to $x = -1$. The horizontal asymptote occurs when the values of the function come close to a single value as the *x*-values approach positive or negative infinity. This rational function is made up of two linear functions, $2x - 5$ and $x + 1$. What happens to the value of this function as the *x*-values grow toward positive infinity? As the *x*-values get larger and larger, the constants subtracted and added have smaller and smaller impact, making the ratio (hence the term *ratio*-nal function) of these two expressions dependent more and more on the $2x$ and the x terms. As this occurs, the ratio is more and more reliant on the x terms, so the ratio of the expressions gets closer and closer to the ratio of the leading coefficients, or $2/1$. The value of this function gets closer and closer to 2, but does not quite reach it. This is also true for the *x*-values leading to negative infinity. For this reason, this function has a horizontal asymptote of $y = 2$.

The *y*-intercept can easily be determined by substituting 0 for x, which would yield $y = -5$. Therefore, the *y*-intercept for this function would be (0, –5). The *x*-intercept can be calculated by finding the value of x that would make the function (*y*-value) equal to 0. Here, by setting the numerator equal to 0, you would accomplish this goal, so $x = 5/2$ is the *x*-intercept for this function. Note that here you rely on the fact that $2x - 5 = 0$ is a linear equation and as such would only have one solution, meaning that the graph can only cross the *x*-axis once. Plotting the intercepts and including the asymptotes provides the graph in figure 4.20.

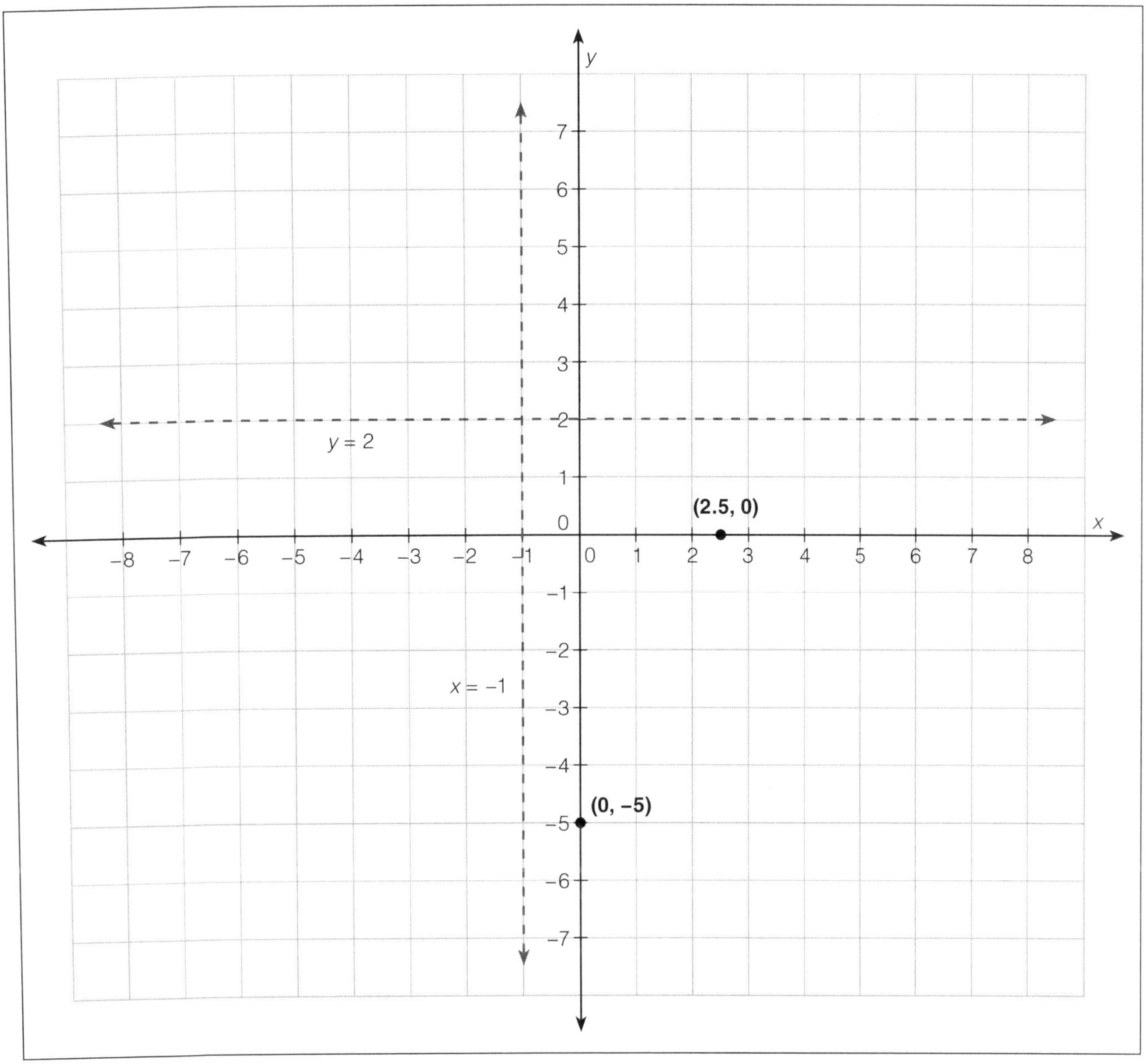

Figure 4.20: Key features of the graph of the rational function $f(x) = \frac{2x - 5}{x + 1}$.

Knowing that the graph will approach the asymptotes, you can sketch in the part of the graph that includes intercepts. By testing a few x-values that are less than –1, you can learn where the remainder of the graph should be drawn. For example, when $x = -4$, the y-value is $\frac{-13}{-3}$, or $4\frac{1}{3}$, and when $x = -7$, $y = \frac{-19}{-6}$, or $3\frac{1}{6}$. This provides information about where the remainder of the graph should be drawn (see figure 4.21, page 96).

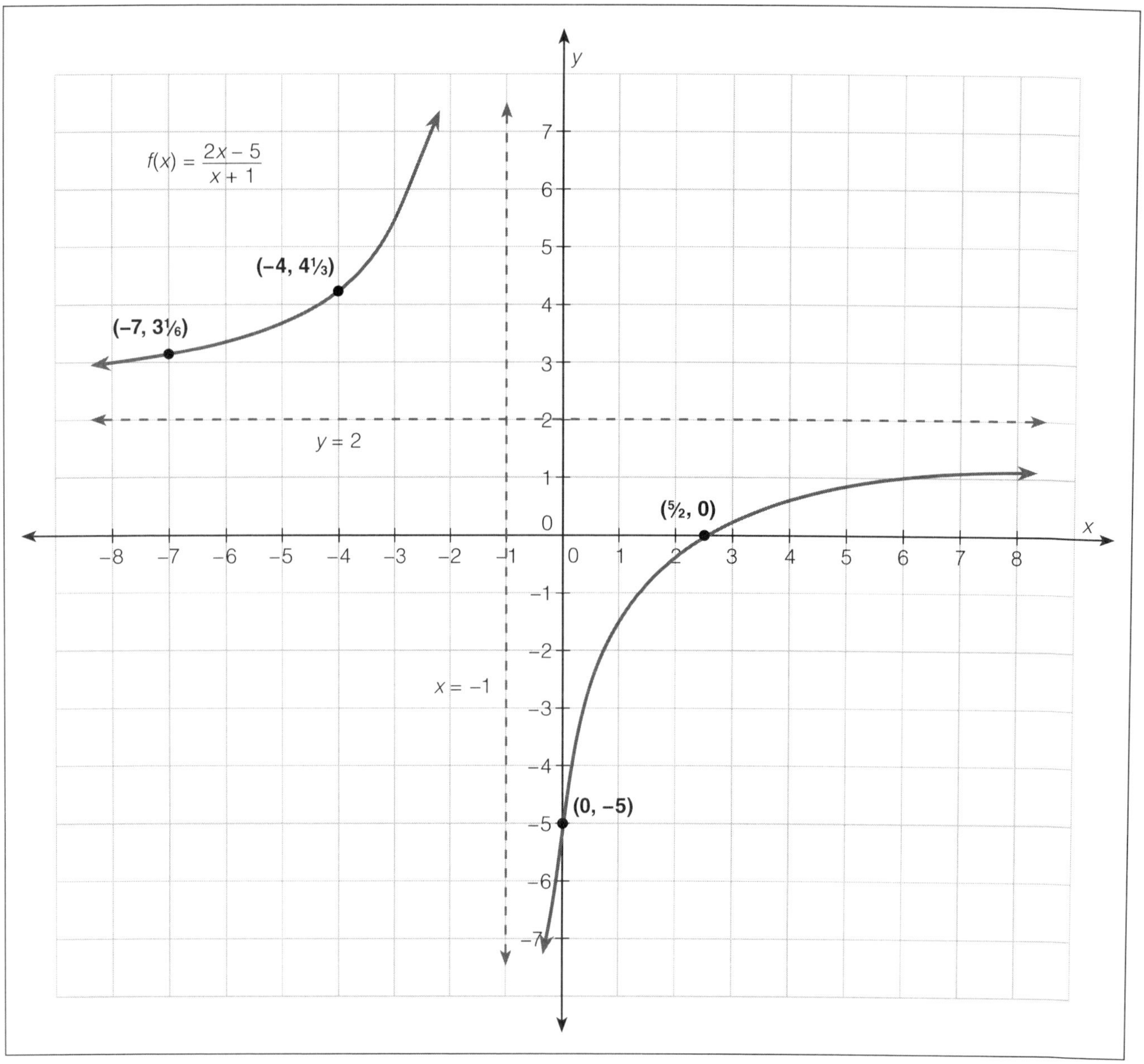

Figure 4.21: Sketch of the rational function $f(x) = \frac{2x - 5}{x + 1}$.

Previously, you explored the role of inverse functions and the manner in which the *x*- and *y*-terms are inverted. How would you develop the inverse function to $f(x) = \frac{2x-5}{x+1}$? Unpacking the function in figure 4.19 (page 94) helps to anticipate the inverse function to $f(x) = \frac{2x-5}{x+1}$. Our original rational function had a domain of $(-\infty, -1), (-1, \infty)$ and a range of $(-\infty, 2), (2, \infty)$. In considering how to create the inverse function, think about the role of the domain and range moving from the original function to the inverse function. Since the domain and range switched places, the inverse function should have a domain of $(-\infty, 2), (2, \infty)$ and a range of $(-\infty, -1), (-1, \infty)$.

Next, consider the vertical and horizontal asymptotes for the original rational function. The original function has $x = -1$ as the vertical asymptote and $y = 2$ as the horizontal asymptote, so the inverse function needs to invert these two asymptotes. Therefore, for the inverse function, you will have a vertical asymptote of $x = 2$ and a horizontal asymptote of $y = -1$. You have observed how inverse functions demonstrate symmetry due to the structure of the Cartesian coordinate system. As the role of the x and y variables are switched in inverse functions, the role of the vertical and horizontal aspects are also interchanged. The value of the vertical asymptote for the original function becomes the value of the horizontal asymptote in the inverse function.

As a final step, look at the intercepts and invert these points to create a sketch of the graph of the inverse function. Whereas in the original graph you had an x-intercept of $(5/2, 0)$ and a y-intercept of $(0, -5)$, the inverse function should possess a y-intercept of $(0, 5/2)$ and an x-intercept of $(-5, 0)$ (see table 4.3).

Table 4.3: Key Information Regarding the Rational Function and Its Inverse

Key Characteristics	
$f(x) = \frac{2x-5}{x+1}$:	The inverse function of $f(x) = \frac{2x-5}{x+1}$:
Domain: $(-\infty, -1), (-1, \infty)$ Range: $(-\infty, 2), (2, \infty)$	Domain: $(-\infty, 2), (2, \infty)$ Range: $(-\infty, -1), (-1, \infty)$
Vertical asymptote: $x = -1$ Horizontal asymptote: $y = 2$	Vertical asymptote: $x = 2$ Horizontal asymptote: $y = -1$
x-intercept: $(5/2, 0)$ y-intercept: $(0, -5)$	x-intercept: $(-5, 0)$ y-intercept: $(0, 5/2)$

How do you use this information to create a graph of the inverse function? Consider how the domain and range, along with the intercepts and asymptotes, all come together to provide a sense of the behavior of this new graph. What do you predict for the algebraic representation of the inverse function? Figure 4.22 (page 98) provides the graph of the inverse function described in the task in figure 4.19 (page 94).

The graph in figure 4.22 should confirm your findings. The exploration to connect a rational function and its inverse function arrived at the function $g(x) = \frac{-x-5}{x-2}$ as the inverse of $f(x) = \frac{2x-5}{x+1}$. The goal of this exploration is to connect earlier mathematical ideas with new ideas to determine inverse functions. Creating mathematical connections between key characteristics of functions such as domain and range, intercepts, and asymptotes allows for explaining the process of creating inverse functions and builds a foundation for future mathematics courses.

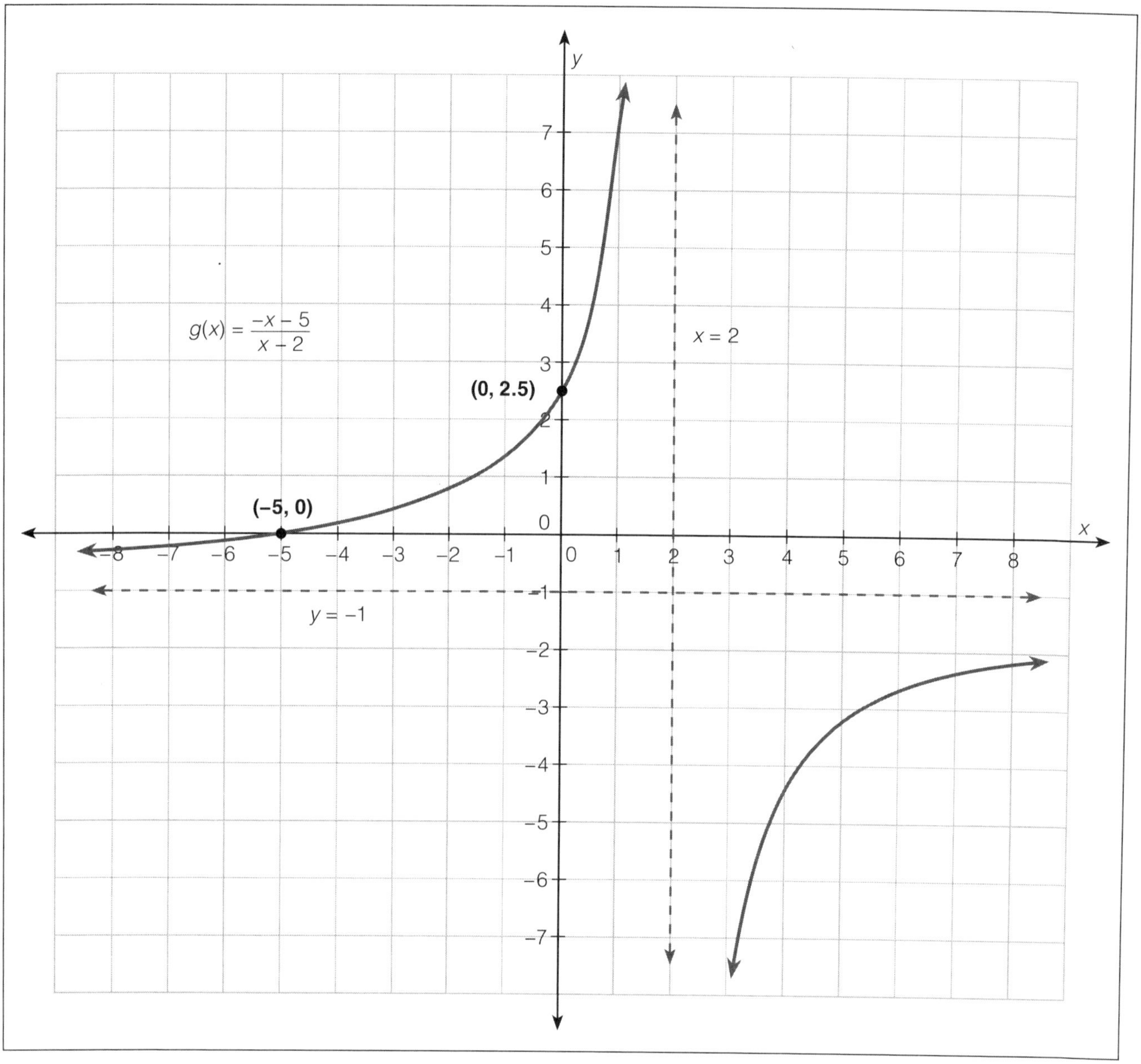

Figure 4.22: Graph of the inverse of the rational function.

The Classroom

You have explored aspects of functions and re-examined your own thinking about connections between various function types. Now, you will have the opportunity to explore what this looks like in a classroom setting. The following videos provide perspective and valuable insights into student thinking and student misconceptions regarding different function types. In the first video, students are engaged with a task to determine different types of functions from multiple sets of points on a graph. You are strongly encouraged to watch the video in its entirety prior to proceeding.

www.solution-tree.com/Comparing_Types_of_Functions_Graphically

Now that you have seen the video related to different function types, focus on the manner in which students are engaged in making sense of the task. What is the role of small-group discussion in informing whole-class discussion? How does the teacher facilitate the discussion through the questions posed?

Perhaps you noticed that the teacher orchestrates discussion rather than providing explicit answers to indicate whether a student is correct or not. The teacher purposefully guides the conversation toward the learning goal of identifying characteristics of different function types based on the responses and contributions of the students. While the teacher engages in the formative assessment process of determining the extent of each student's understanding, he is also gaining a sense of the understanding of the small group and the progress of the whole class related to the topic at hand.

How are students held accountable to listen, examine, and reflect on each other's thinking? At times, the teacher asks one student, "What do you think?" and follows that up by asking other students, "How do you know?" As the conversation continues, the questions evolve from *what* any one particular student knows to questions about *how* and *why* students know to shed light on the manner in which the students are thinking about the mathematics involved.

In asking about the points that illustrate a linear function type, the teacher guides the discussion related to the rate of change of linear functions. The teacher is able to elicit student responses that address the behavior of the groups of points provided. Next, with the points that indicate a quadratic relationship, the teacher facilitates the small-group discussion to not only identify the points, but also to explain the relationship between the points. The students are asked to reason numerically with (–1, 1), (1, 1), (–2, 4) and (2, 4), and then they are challenged to see whether the quadratic relationship could be generalized for all the points when the teacher asks, "Does that work for all?"

As seen in this video, students are expected to regularly engage in Mathematical Practice 3, "Construct viable arguments and critique the reasoning of others," so that this practice becomes the standard operating procedure in the classroom. While it might be easier to focus attention on individual students, learning needs to be a social endeavor during which students benefit from each other's perspectives and are held accountable to understand both their own reasoning and the reasoning of others. Since the Mathematical Practices are consistent throughout the various grade bands, students become accustomed to examining multiple perspectives and refining their own mathematical thought processes by constructing valid arguments. As such, Mathematical Practice 3 can and should be at the heart of many classroom conversations.

What do you notice when the students are asked to reason with the shapes of the quadratic graph? How does the conversation move from juxtaposing the constant rate of change in linear functions to the behavior of the parabola as the points decrease from (–2, 4) to (–1, 1) and then increase from (1, 1) and on to (2, 4)? As illustrated in this video, students often gesture the behavior of the curve to describe the motion, and mention numerical relationships between values they examine. By being asked to examine the specific points and then encouraged to generalize that exploration to behavior of functions that demonstrate a quadratic function, students are engaged in Mathematical Practice 7, "Look for and make use of structure." The relationship explored numerically with the points provided can and is generalized to the structure of quadratic functions when the student in the video notices that "points are mirrored in a quadratic function." As a teacher, enabling students to reason through specific cases and then generalize the mathematical notions using structure is essential as students transition in their mathematical journey.

In the second video, students are asked to solve $x^2 - 4x = 5$ by considering each side of the equation as a separate function. The teacher instructs the students to think of this equation as $f(x) = g(x)$ and asks them to sketch each function separately. Students are asked to make sketches, talk in small groups, and then use the information to make sense of the problem. Pay particular attention to the sketches on the students' whiteboards prior to the small-group and whole-class discussions.

www.solution-tree.com/Solving_f(x)_=_g(x)_Graphically

How do the purposeful actions of the teacher in asking students to sketch the equations as two separate functions impact the lesson? The students provide many different types of sketches during the lesson, demonstrating to the teacher what students do and do not know about the graphing of quadratic equations—a quick and helpful addition to the formative assessment process.

What did you notice regarding student misconceptions? Perhaps you noticed that many of the students recognize that the quadratic function should look like a parabola. However, some students draw the parabola opening downward. As evident by their sketches on their whiteboards, some students correctly factor $x^2 - 4x$ as $x(x - 4)$ and indicate that 4 should play a role in the graph. However, many students incorrectly shift the parent graph of $y = x^2$ four units to the *left* or four units *up*. This lesson provides you the opportunity to consider what actions you take as a teacher when students represent misconceptions as they are engaged in a lesson.

TQE Process

At this point, it may be helpful to watch the second video again and pay close attention to the tasks, questioning, and opportunities to collect evidence of student learning.

The TQE process can help you frame your observations. Teachers who have a deep understanding of the mathematics they teach:

- Select appropriate *tasks* to support identified learning goals
- Facilitate productive *questioning* during instruction to engage students in Mathematical Practices
- Collect and use student *evidence* in the formative assessment process during instruction

The *task* chosen for this lesson involves solving a quadratic equation by graphically examining each side of the equation as a separate function—namely $f(x) = g(x)$. The learning goal is for students to understand that the number of solutions can be anticipated by examining each side of the equation, and then the actual solutions can be determined by reasoning numerically, graphically, and algebraically. The task supports this learning goal by allowing students opportunities to explore a quadratic function $f(x)$ and a constant function $g(x)$ by representing them graphically. Students draw the graphs and make conjectures. They then compare, share, and discuss their solutions and make connections between the graph and the original equation. In doing so, they are more likely to engage in a productive struggle that leads them to experience dissonance, which entices them to examine the problem with deeper insights. This aspect of the task is important to achieving the learning goal of this lesson.

The *questioning* by the teacher creates a classroom atmosphere that welcomes student participation and interaction. When the teacher asks, "Who would like to share their graph with us?" the students are invited and encouraged to participate in the lesson. By providing a voice to various students in classroom discussion and pointing to what students have written on their whiteboards, the teacher holds students accountable and asks them to "show us what you have done." In addition, the teacher remains focused on the learning goal of the lesson by asking, "How many solutions do we have to this equation?" A student responds by indicating that the line would cross the parabola in two spots and so there would be two solutions. Notice that the teacher maintains the goal of the problem by asking, "Do there have to be two solutions?" The student responds, "There does not always have to be two solutions." The teacher includes another student by asking, "Do you agree with that? Why?" Here, the teacher is encouraging students to be engaged in Mathematical Practice 3, "Construct viable arguments and critique the reasoning of others."

The teacher's use of *evidence* of student thinking and understanding provides him information to navigate the classroom discussion. The teacher walks around the room and examines what different students are sketching and what algebraic steps they have written on their whiteboards. The teacher can quickly determine which students are struggling with graphs and which students understand graphs but not the idea behind finding points of intersection. The teacher is collecting information on how the students are developing understanding of the specific learning goal of determining the number of solutions in solving quadratic equations. Note how he proceeds to push for more precision while promoting Mathematical Practice 8, "Look for and express regularity in repeated reasoning." The teacher asks, "What would have to happen for there to be *no* solution?" and elicits the response from a student that $g(x)$ (currently $g(x) = 5$) would have to be a negative number. In asking "Could it be *any* negative number?" the teacher pushes for precision, and the student rises to the challenge by providing further evidence that "no, the $g(x)$ would need to be a negative number below the parabola in order for you to have no solutions." The student expands on the original response by saying, "There is no point of intersection between the line and the parabola." The teacher pushes to collect the evidence that is needed to determine whether the student has met the learning goal.

The Response

Students often see mathematics as a disconnected series of procedures without having meaningful opportunities to engage in tasks that focus on concepts. During classroom instruction, you have the opportunity to develop student reasoning and sense making while engaging students in various Mathematical Practices that support learning. Many times, students expect that solving algebraic equations should only involve the process of manipulating and isolating variables. While that procedure will yield numerical solutions to equations, it is not sufficient to provide students the understanding of the meaning of that solution or the broader sense of the solution using other representations. Students need to experience tasks that demonstrate the value of using different representations as well as how these representations or various forms of an equation provide different insights into solving both contextual and mathematical problems.

In addition to wanting to solve problems through algebraic manipulation only, students often struggle with understanding the correct role of parameters in the families of different functions. This is particularly evident in horizontal translations, where the shift often feels like the reverse of what is represented in the function. For example, in the function $y = (x - 3)^2$, many students will want to shift the parent function of $y = x^2$ three units to the left to represent the –3 horizontal translation, rather than making the shift three units to the right. This is where the conceptual approach to making sense of the family of functions is an important factor in overcoming common misconceptions. Consider the value for which the expression $(x - 3)$ is equal to 0, and how that value of 3 will indicate the correct direction of the translation. Teachers need to provide opportunities for students to understand transformations numerically, graphically, and algebraically, using many different representations to solve problems, allowing students to see the how the transformations relate to one another. Students need to understand how the different representations work together to provide insight into how to represent functions correctly, and this can be developed through exposure to rich and engaging problems.

Reflections

1. What do you feel are the key points in this chapter?
2. What challenges might you face when implementing the key ideas for the development of understanding of different types of functions? How will you overcome them?
3. What are the important features for developing an understanding of different types of functions, and how will you ensure your instruction embeds the support needed for these features?
4. Select a recent lesson you have taught or observed focused on different types of functions. Relate this lesson to the TQE process.
5. What changes will you make to your planning and instruction based on what you read and considered from this chapter?

CHAPTER 5

Modeling With Functions

How does the understanding of various function types enable you to make sense of and mathematize the world around you? This chapter explores ways to integrate the understanding of functions and algebraic concepts to make sense of the world around you. Real-world situations provide the setting to investigate functions and to purposefully use mathematics to explain, interpret, and make predictions through the use of modeling with functions.

The Challenge

A critical part of learning mathematics involves connecting mathematical content to relevant contexts in order to explore and analyze situations. Mathematical Practice 4, "Model with mathematics," focuses on this aspect of mathematizing the world by using mathematics to examine, predict, and gain greater insight into everyday phenomena. What are some different ways that you can make sense of data using the mathematics studied in high school? Consider the task in figure 5.1.

DO NOW

How would you model the number of Facebook users since 2003?

Years After 2003	Number of Users (in Millions)
1	1
2	5.5
3	12
3.4	20
3.8	50
4.8	100

Years After 2003	Number of Users (in Millions)
5.1	150
5.4	200
5.7	250
6	350
6.2	400
6.7	500
7	608

Years After 2003	Number of Users (in Millions)
7.7	750
7.9	800
8	845
8.3	901
8.6	955
8.7	1010
9.3	1110

Sources: Associated Press, 2012; Facebook, 2013.

Figure 5.1: Modeling the number of Facebook users task.

How did you begin to think about how to model the number of Facebook users? You may have started by looking at the first year and the last year of the data provided and thinking about how a linear relationship might model these data. Or, perhaps you started with the change in the number of users during the first few years and looked to see if a pattern emerged. What impact would the differences in the numbers of users for consecutive data points have on your analysis? You may have explored to see if the rate of change in the number of users remained the same throughout the years that are represented in the table. Consider the differences between consecutive entries for the number of users as shown in table 5.1.

Table 5.1: Differences in the Number of Facebook Users

Years After 2003	Number of Users (in Millions)	Difference From Previous Number of Users
1	1	
2	5.5	4.5
3	12	6.5
3.4	20	8
3.8	50	30
4.8	100	50
5.1	150	50
5.4	200	50
5.7	250	50
6	350	100
6.2	400	50
6.7	500	100
7	608	108

How do knowing consecutive differences in the entries for the number of users help you think about what function to use as a model for the number of Facebook users? You may be thinking about the connection between arithmetic sequences and linear functions, where the pattern from arithmetic sequences is represented as differences in consecutive values in the dependent variable. In arithmetic sequences, the terms (the ordered numbers in the sequence) increase by a constant rate (which is why there is a link to linear functions, as they have a constant rate of change). For example, the sequence {0, 50, 100, 150, 200, 250} is an arithmetic sequence with a first term of 0 (represented as $a_1 = 0$). You see that this sequence is embedded in the number of users given. How can you make sense of this pattern? The recursive representation—where you can build the sequence one step at a time—is $a_n = a_{n-1} + 50$, where n is the ordered location of the number in the sequence, a_n is the value of the number of the term in the nth place, and a_{n-1} is the value of the term immediately prior to the nth term, or the $(n - 1)^{th}$ term. The explicit representation—where you can get to any number in the sequence regardless of whether you know the value of the previous term—is $a_n = a_1 + d\,(n - 1) = 0 + 50\,(n - 1)$, where d is the constant difference between consecutive terms. How does this representation connect to linear functions? As arithmetic sequences are built using a constant rate of change, they represent linear functions. You can rewrite the explicit representation in the form $y = mx + b$ as $a_n = 50n - 50$. When we think of the number of the term of the sequence as the independent variable and the value of the term as the dependent variable, the 50 that multiplies n is the rate of change and the –50 (which is determined by $a_1 - 50$, or $0 - 50$) is the y-intercept (which would connect to the "zeroth" term, or a_0, if such a term existed).

What is interesting about the information in table 5.1 is that if you think of the number of Facebook users as the dependent variable (and number of years after 2003 as the independent variable), there is an arithmetic sequence as part of the data set. However, what you need to be cautious of with this data set is whether there is a pattern in the independent variable that would represent a consistent interval for the terms of the sequence. For most of the data set, the interval between the number of years is not constant. As you see in the number of years since 2003, there is a growth pattern of 0.3 years between 4.8 and 6 (4.8, 5.1, 5.4, 5.7, and 6), which closely corresponds with the constant rate of growth of 50 million users per time interval. While these data points will form a linear relationship, you may not be certain whether this will work for most of the data points. However, note that the last value (6, 350) does not follow the arithmetic growth pattern. You probably also see that the interval in the number of years after 2003 is not constant for most of this data set, which makes the analysis of whether a linear model is appropriate for the entire data set difficult to determine. You see that a linear function will model one part of the data set, but from the analysis of the table, it is difficult to see whether a linear function is best for the entire data set.

What other methods are available to analyze data to help you consider other types of functions to use as models? How can various patterns in the differences between the number of users be explored? Examine the population of a small town over a twelve-year period (see table 5.2).

Table 5.2: A Small Town's Population

Year	Population
1	120
2	135
3	160
4	195
5	240
6	295
7	360
8	435
9	520
10	615
11	720
12	835

You may notice that while there is a constant difference in the independent variable (year), the difference between consecutive terms of the population is not constant; therefore, this is not a linear relationship. But is there a pattern in the differences (see table 5.3)?

Table 5.3: Differences of a Small Town's Population

Year	Population	Difference From Previous Year
1	120	
2	135	15
3	160	25
4	195	35
5	240	45
6	295	55
7	360	65
8	435	75
9	520	85
10	615	95
11	720	105
12	835	115

While the differences are not constant, there is a pattern to the growth. Each year, the rate of growth increases by ten more people than the previous change. In fact, the rate of change of the population each year has a consistent rate of growth—note that the second differences are constant. Or, stated a little differently, the rate of change has a constant rate of change (see table 5.4).

Here, you see a quadratic relationship, one that is represented by a polynomial function with the greatest power of 2. One way to categorize polynomials involves examining the number of layers of common differences it would take to arrive at a common difference, or constant rate of change. If a polynomial demonstrates a constant rate of change in the first differences between consecutive terms of the dependent variable, the relationship between the points is linear. If the first rate of change does not demonstrate a constant difference but the set of second differences is constant, then the polynomial is not linear but would be classified as quadratic. In fact, the number of differences it takes to get to a constant rate of change helps identify the degree of the polynomial (which students explore further when they get to calculus).

Given the original table of values in figure 5.1 (page 103), how would you respond to a student who examined first and second differences as provided in table 5.5?

Table 5.4: Differences in the Differences of a Small Town's Population

Year	Population	Difference From Previous Year	Difference in the Differences
1	120		
2	135	15	
3	160	25	10
4	195	35	10
5	240	45	10
6	295	55	10
7	360	65	10
8	435	75	10
9	520	85	10
10	615	95	10
11	720	105	10
12	835	115	10

Table 5.5: First Differences and Second Differences in Number of Facebook Users

Years After 2003	Number of Users (in Millions)	First Difference	Second Difference
1	1		
2	5.5	4.5	
3	12	6.5	2
3.4	20	8	1.5
3.8	50	30	22
4.8	100	50	20
5.1	150	50	0
5.4	200	50	0
5.7	250	50	0
6	350	100	50
6.2	400	50	–50
6.7	500	100	50
7	608	108	8

The student may see that there is not a constant rate of change in the second set of differences, so this data set does not appear to be a good candidate for a quadratic model. You should also consider how to support the student to understand that analyzing the data set in this manner is not an appropriate technique as the interval in the number of years after 2003—other than in the interval that is already modeled well by a linear function—is not constant. At this point, it is a challenge to determine if any polynomial functions will be a good fit for the data.

What other functions can be examined by looking just at the data expressed in a table? In order to think about other functions that may be appropriate, consider data based on investing an amount of money over a period of time. Start with an initial investment of $125 and look at how that amount of money you invested would change at the end of each year over eight years (see table 5.6).

Table 5.6: Amount of Money Invested Over Eight Years, With Differences Between Years

Year	Amount of Money	Difference From Previous Year
1	125.00	
2	135.00	10.00
3	145.80	10.80
4	157.46	11.66
5	170.06	12.60
6	183.67	13.61
7	198.36	14.69
8	214.23	15.87

You see that this is not a linear pattern, as the rate of change is not constant. In fact, as you can see in table 5.7, there is not a constant rate of change in either the second or the third differences, so it appears that a polynomial model is not a good fit for this data set.

Table 5.7: Amount of Money Over Eight Years, With Differences Between Years

Year	Amount of Money	First Difference	Second Difference	Third Difference
1	125.00			
2	135.00	10.00		
3	145.80	10.80	0.80	
4	157.46	11.66	0.86	0.06
5	170.06	12.60	0.94	0.08

Year	Amount of Money	First Difference	Second Difference	Third Difference
6	183.67	13.61	1.01	0.07
7	198.36	14.69	1.08	0.07
8	214.23	15.87	1.18	0.10

How does the context of this data set, investing money over time, present a different way to consider how to analyze data? How do you determine the next year's value of an investment? In calculating a simple interest growth over time, you multiply using an interest rate in order to determine the next year's value of money. An explicit representation of this over any number of years is $y = A(1 + r)^x$, where y is the total amount of money at a given time, A is the original amount of money, r is the simple annual interest rate, and x is the number of years. This is an example of an exponential function. How would you analyze a data set to determine whether it follows an exponential model (see table 5.8)?

Table 5.8: Amount of Money Invested Over Eight Years, With Differences Between Years

Year	Amount of Money Invested	Ratio of Current Year to Previous Year
1	125.00	
2	135.00	1.08
3	145.80	1.08
4	157.46	≈1.08
5	170.06	≈1.08
6	183.67	≈1.08
7	198.36	≈1.08
8	214.23	≈1.08

Since you multiply by a constant value in an exponential function, the consecutive values in the dependent variable (in this example, the amount of money invested by year) should have a constant ratio between them. As you can see from the information in table 5.8, the data set demonstrates a constant ratio between the terms (with slight adjustments needed because of rounding to the nearest cent), showing that this is a method of data analysis you can use to determine an exponential relationship. The explicit rule modeled by this data set would be $y = 125(1 + 0.08)^x$.

How does this analysis correspond to the number of Facebook users data set? Exponential function analysis is predicated on consistent differences in the independent variable. Since much of the data for the number of Facebook users is provided in a way that precludes the use of examining constant differences, you may need to explore other techniques to model the data. You may consider extending the analysis to look at these values graphically. Consider the graph of all of the data points (see figure 5.2, page 110).

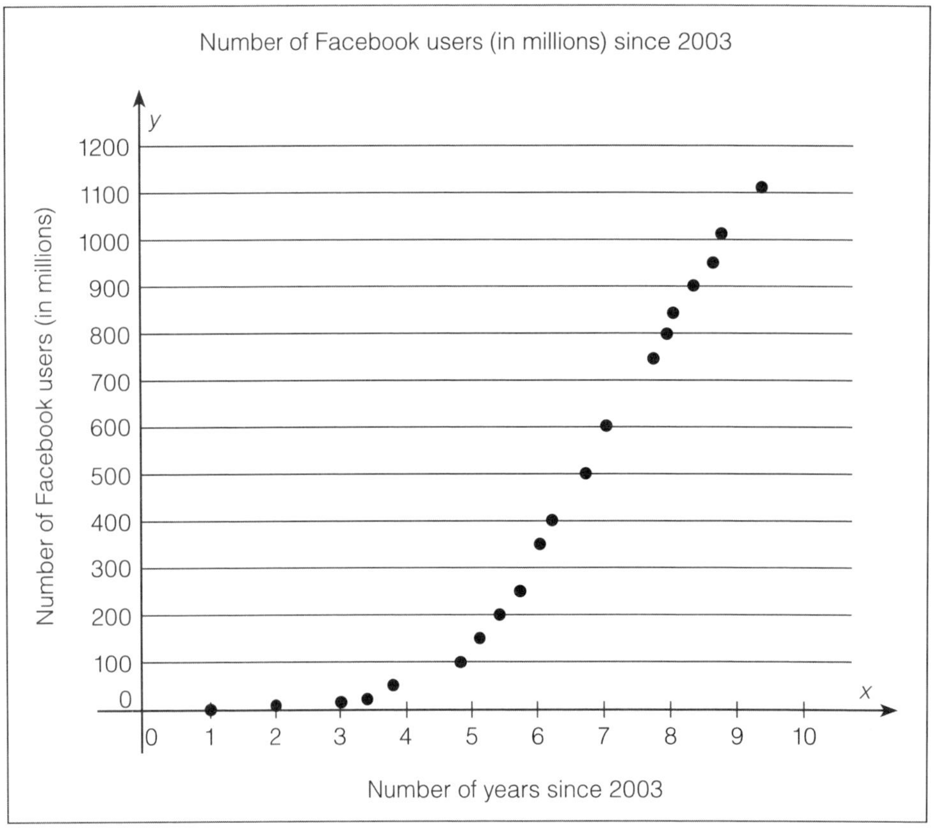

Figure 5.2: Graphical representation of the number of Facebook users.

Through a quick visual examination of the number of Facebook users in figure 5.2, what types of functions come to mind? You see how the different patterns that you explored numerically appear in the graph. Looking at the series of points, you see that they certainly seem to start off rather slowly and then begin to rise rather quickly at about five years after 2003. You may feel that the data set of the number of Facebook users demonstrates different function behaviors over different intervals of time—some places may look like linear functions, other places like quadratic or exponential functions. How do you address a data set that exemplifies different function behaviors? We will explore this question later in the chapter.

This task provides an example of how you examine, analyze, and reason while modeling with mathematics. Modeling is a major focus of high school mathematics and beyond. The choice of models needs to be influenced by the situation that you are analyzing. What role did the context play in informing your approach to this task? What tools and representations enable you to explore the different mathematical functions that may be needed to successfully make predictions? How do you gain a deeper insight to making sense of the mathematics in a coherent and analytical fashion through modeling?

In this chapter, these questions will be addressed as you continue to consider the significance of context in reasoning and sense making while mathematizing real-world situations. You will have opportunities to explore the manner in which a deep understanding of function types allows for multiple ways to model situations. Conceptual understanding will continue to be emphasized, along with the strategies and procedures that support thinking and assist in arriving at multiple ways to model with mathematics.

The Progression

As explored in chapter 4, students extend their understanding of functions throughout high school by analyzing characteristics of functions using multiple representations. They investigate quadratic, exponential, and logarithmic functions as well as radical, rational, and trigonometric functions. The following progression shows that developing an understanding of functions in algebra entails the analysis of *when*, *why*, and *how* to use each of these function types to mathematize the world.

- Analyze functions of different types using multiple representations linked to context.
- Determine and justify the selection of a function type to model a situation.
- Explore and extend the notion of rate of change with different function types.
- Compare and interpret functions in terms of the situation they model.
- Construct models to represent situations and make predictions.

High School

One goal of high school mathematics is to support the ability to create functions that make sense of, model, and predict outcomes of real-world situations. It is crucial to possess a deep understanding of different function types involving numeric, graphic, and algebraic representations of function behavior. By understanding different characteristics of function types, such as their domain, range, rate of change, and asymptotes, students determine which type of function would be best to model a particular situation. Specifically, as it pertains to the notion of rate of change, consider how students experience the progression from linear functions to quadratic functions to exponential and other function types. Quite intentionally, students first explore constant versus nonconstant rates of change as they transition from middle school to high school. Next, they are led to extend the ideas explored through analyzing slopes of lines (hence the term *linear* functions) to rate of change as a concept that helps to classify and distinguish various function types. Gradually, students explore the rate of change of various functions as exemplified by quadratic, exponential, trigonometric, and rational functions and gain an understanding of how different functions change values and the implications of these changes numerically, graphically, and analytically. In fact, on the mathematical horizon for many students are courses such as precalculus and calculus that delve into rate of change using more complex mathematical analysis. The progression of topics allows for ways to analytically explore, calculate, and literally *differentiate* how, when, and why functions vary in different intervals within the study of differential calculus. This corresponds to a first-year calculus course primarily focused on a deep understanding of the concept of rate of change in various function types.

More specifically, within the progression, the designation of mathematical topics incrementally extends and generalizes the idea of slope from $^{\text{Rise}}/_{\text{Run}}$ to $^{\text{Change in } y}/_{\text{Change in } x}$ to $^{\Delta y}/_{\Delta x}$ (delta y over delta x). As students gain mathematical maturity through their algebra and geometry experiences, the designation of $^{\Delta y}/_{\Delta x}$ using the Greek letter delta is replaced with $^{dy}/_{dx}$ to represent the rate of change of y with respect to x. Note the purposeful use of the letter *d* for the connection to differentials and derivatives in calculus (see figure 5.3, page 112).

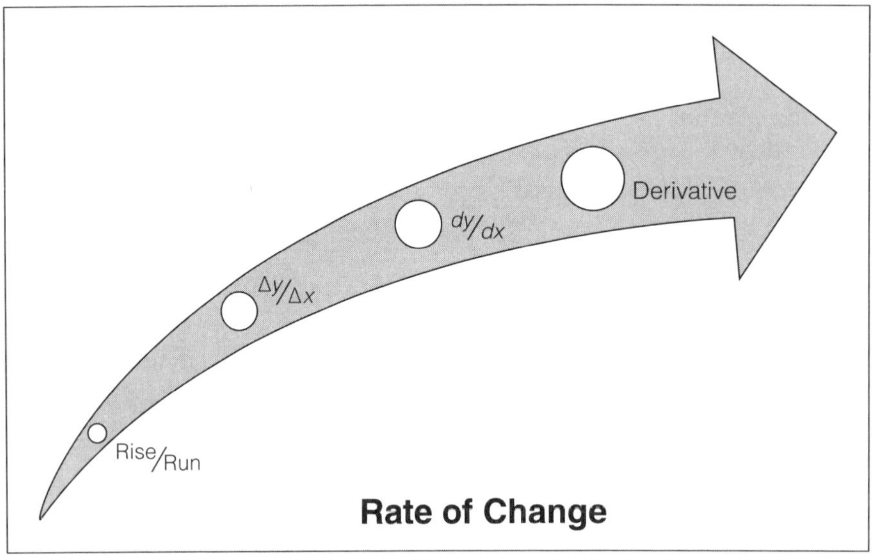

Figure 5.3: Progression of ideas related to rate of change.

Throughout this entire progression of ideas, the central mathematical concept remains the notion of rate of change—explored through multiple representations with various function types. It is this connected and incremental development of a mathematical concept that illustrates the significance of the need for a conceptual understanding of mathematics throughout the progression.

The Mathematics

Through analysis of situations and making sense of different representations, students have the opportunity to determine appropriate models for given contexts. They learn to mathematize situations. This is described in the Mathematical Practices as modeling with mathematics. High school students must be able to model with quadratic equations, exponential functions, radical functions, and piecewise functions. They benefit from varied experiences exploring contexts and supporting their decisions in determining the best models to represent those contexts mathematically.

Modeling With Quadratic Functions

Returning to the Facebook users example (see table 5.1, page 104), you already explored some function types as potential models using numerical data. Reconsider the task using graphical representations. How would various function types model this situation from a graphical perspective? Considering the curvature illustrated by the points provided in the graph in figure 5.2 (page 110), would a quadratic function appropriately model the total data set of the number of Facebook users? How would you explore possible quadratic functions that might be used to model the situation? This is where you can consider the strategic use of technology to help you model a function that represents a majority of the data points. Consider entering the points into a table and then graphing them using technology. Many forms of technology will allow you to overlay a quadratic function with changeable parameters, such as $f(x) = ax^2 + bx + c$. Before you begin to explore, however, think about the impact of parameters discussed in chapter 4. What about the value of the parameter a? Because the curve opens "up," the value of a needs to be positive. Now you can explore different values for a, b, and c that will model the function. With some persistence in refining the model by modifying the parameters in an iterative process, you can create an increasingly precise match of the data. You may come up with a model similar to the one in figure 5.4.

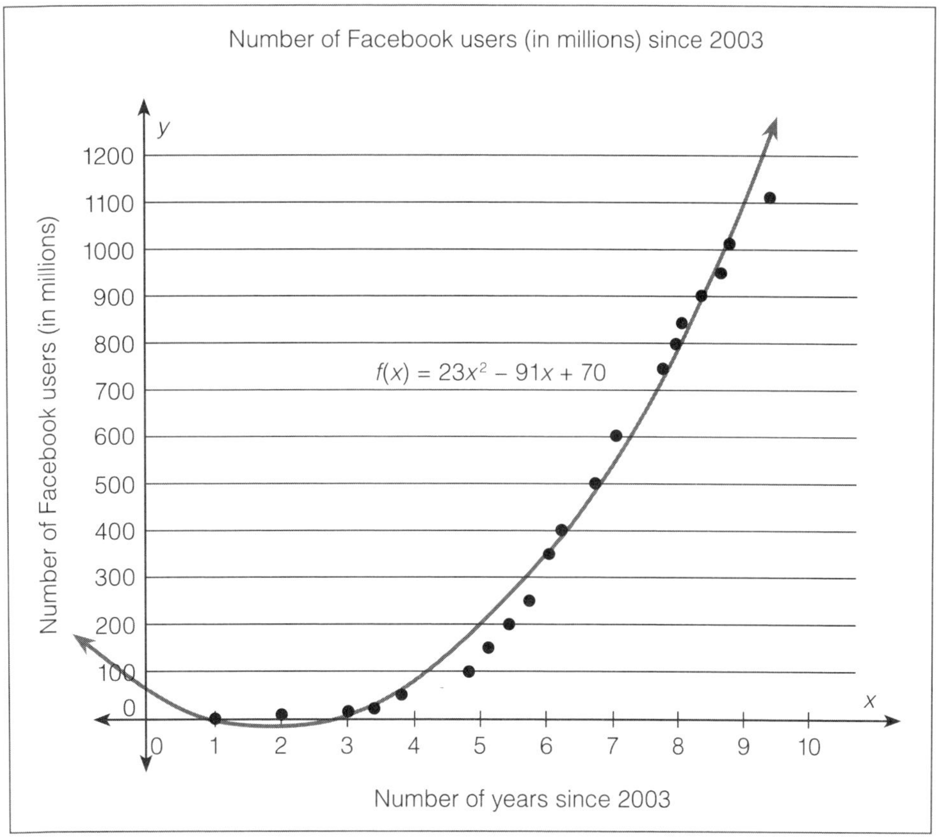

Figure 5.4: Quadratic modeling of Facebook users.

As you can see from the graph, this model is a good fit, but there are still places where the quadratic function is not a close fit to the actual data. What is important in this type of exploration is how you are making sense of adjusting the parameters to create a model to fit the data. The exploration becomes a sense-making activity, not merely a computational one. Different explorations may create slightly different models, based on your decision about where to have the best fit for the function. The function model in figure 5.4 is the quadratic equation $f(x) = 23x^2 - 91x + 70$, where x refers to the number of years after 2003 and $f(x)$ represents the number of Facebook users in millions. How well does this model represent the data, particularly for the first four years? Notice that the model shows that the function values would be negative between years one and three. According to the context, this would represent a negative number of Facebook users—which indicates that this quadratic model is not appropriate for the entire graph. This is a place where it is important for you to make connections between the context of the task and the model you choose to represent that context.

How well do the values of the quadratic curve model data points on this interval from $x = 4$ to $x = 6$? You notice that the quadratic curve is above the actual data values. Higher y-values represented by the quadratic curve show that the equation provides an *overestimate* on this interval as they occur above the actual data points. How well does this quadratic model represent the context beyond $x = 6$? Where does the quadratic model of $f(x) = 23x^2 - 91x + 70$ represent an *underestimate*? How do you know that? At seven years after 2003, you see that the data point is above the curve, showing that the function is an underestimate. Thinking about where the model overestimates and underestimates the actual data enables students to better understand the role of mathematical modeling in representing the world around them.

Modeling With Exponential Functions

At this point, you may want to explore other types of functions as models for this context. You have looked at linear functions and quadratic functions. How does the rate of increase of the number of Facebook users provide suggestions for another type of function that could model the data? Of particular note in the data is that the number of Facebook users increases slowly in the first several years but then increases much more quickly for a few years. Which function type might best model this aspect of the data? Re-examine the functions explored in chapter 4 where $g(x) = x + 2$, $h(x) = x^2$, and $l(x) = 2^x$ (see figure 4.1, page 75).

Suppose you wanted to describe the average rate of change between common function values as a way of comparing $g(x)$, $h(x)$, and $l(x)$. This perspective of rate of change can be described as how quickly functions arrive at specific, common function values. Consider the function values of 4 and 64 for all three functions. For the linear function $g(x)$, the independent variable values to produce function values of 4 and 64 are $x = 2$ and $x = 62$. For the quadratic function $h(x)$, the independent variable values are $x = 2$ and $x = 8$. The last function, $l(x)$, arrives at its initial targeted value at $x = 2$ and the final targeted function value at $x = 6$. This tells you that $l(x)$, the only exponential function of the three functions, has the greatest average rate of change for the given function values. Exponential functions (with a base greater than 1) are characterized by relatively small rates of change at the start and then gain momentum as the independent values increase. The average rates of change can be described as the slope of the secant line determined by the points identified by the two function values. A *secant* line of a curve is a line that intersects the curve at two points. Figure 5.5 provides an example of the secant lines for $h(x)$ and $l(x)$.

It is clear that the slope of the secant line to a function is determined by, and therefore dependent on, the specific points selected. What happens as those points get closer and closer together? This exploration allows for an extension of the notion of rate of change explored algebraically to the geometrical definition of rate of change that is a central focus in calculus. A broad view of this extension is provided here; however, the details of this analysis are beyond the scope of this book.

As the distance between the two points determining the secant line for these continuous functions becomes infinitesimally small, the slope of the secant line approaches the slope of the tangent line. The tangent line touches the function at just one point and describes the *instantaneous* rate of change at that point. It is different from the *average* rate of change determined by the slope of the secant line because the tangent line describes the rate of change of the function at one specific point rather than at more than one point. Consider the points B and D, where both functions have the value of 64. The slope of the tangent line to $h(x)$ at B is 16, and the slope of the tangent line to $l(x)$ at D is about 44—indicating that the values for $l(x)$ are changing more rapidly than the values of $h(x)$ when $y = 64$. Figure 5.6 provides a graph of the tangent lines at those points.

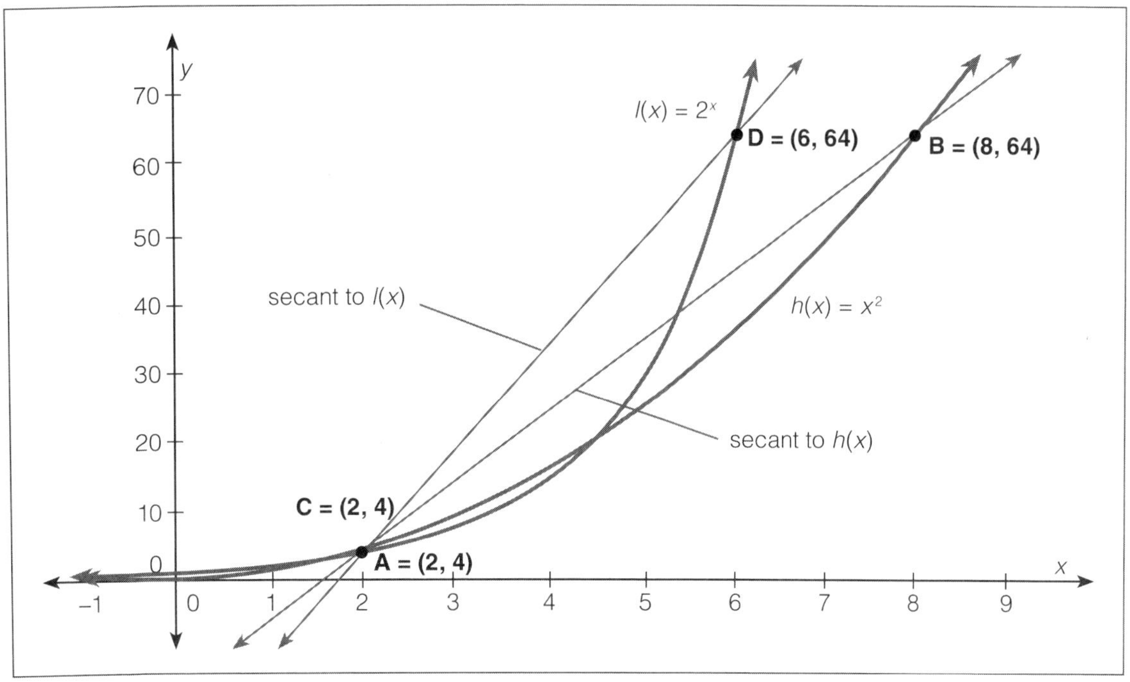

Figure 5.5: Graph of secant lines to $h(x)$ and $l(x)$.

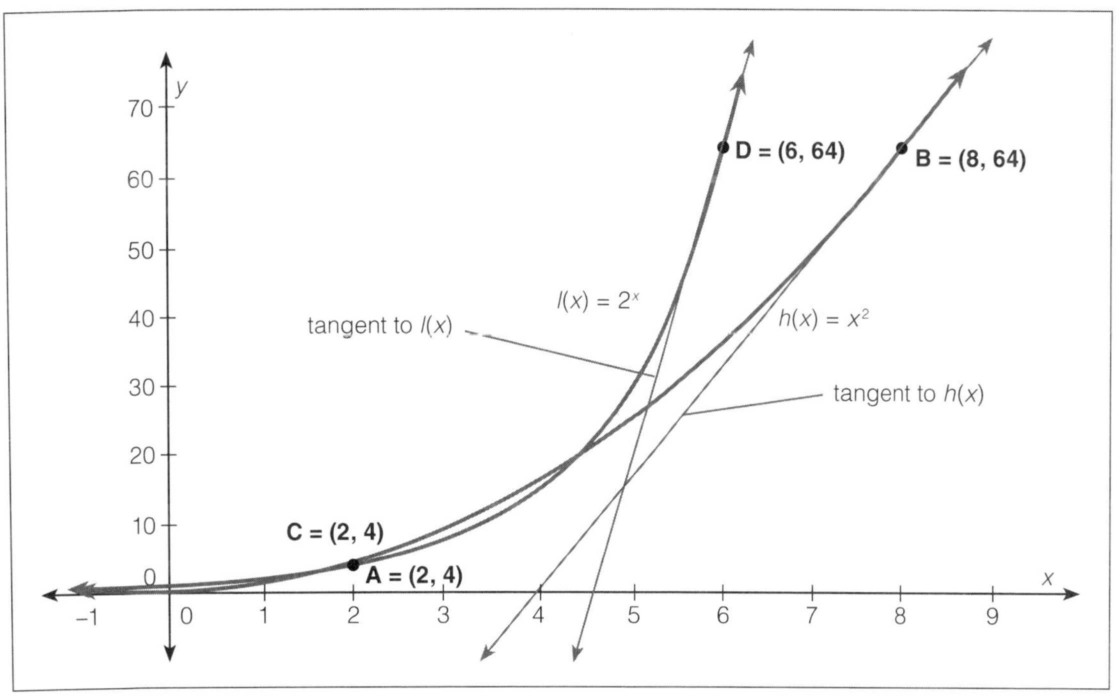

Figure 5.6: Graph of tangent lines to $h(x)$ and $l(x)$ at the function value of 64.

Notice how the exponential function, $l(x)$, becomes steeper at a faster rate than $h(x)$ as x increases. This change in rate of change seems to be a good model for the Facebook users data. Attention then turns to determining an appropriate exponential function.

In what ways can you compare and analyze the rate of change of different exponential functions such as $y = 2^x$, $y = 3^x$, and $y = e^x$? Consider the graphs for $y = 2^x$, $y = 3^x$, and $y = e^x$ in figure 5.7.

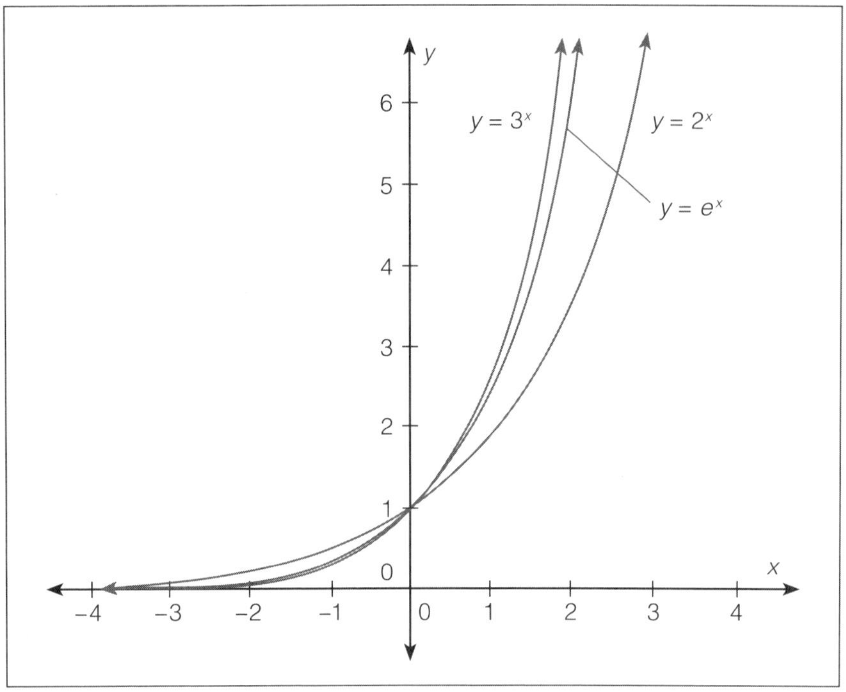

Figure 5.7: Comparing graphs of different exponential functions.

As you can observe from the graphs in figure 5.7, they share the same key point (0, 1), as would all exponential functions that have not been translated. Additionally, all three exponential functions have a base greater than 1 and demonstrate similar behavior with the function values rising dramatically as the x-values increase. However, while they all rise dramatically, the rate of change at each point in the graphs is not the same. Which function's rate of change is increasing at the fastest rate? Visual examination will tell you that the rate of change for $y = 3^x$ is increasing the fastest of the three functions. You could also examine this through the use of a table of values. The average rate of change could also be used to compare the functions in a way similar to how you compared them for $g(x) = x + 2$, $h(x) = x^2$, and $l(x) = 2^x$. As mentioned earlier, another method for comparing rates of change of different functions is to examine just one point and the slope of the tangent line at that point.

Unlike linear functions that display a constant rate of change or quadratics that show a constant rate of the rate of change (as shown in the second differences of table 5.4 on page 107), the exponential function $f(x) = e^x$ has the mathematical distinction of being *its own* rate of change. To analyze the variation of some other function types, you would need to engage in lengthy analysis—and then perhaps arrive at an approximation—to determine the particular rate of change for specific points of a function. With

e as the base of the exponential function, the rate of change is the same as the function value at a given point. This fact allows the exponential function $f(x) = e^x$ to serve a unique role in mathematical modeling.

Return to the exploration related to the number of Facebook users. What do you notice about the use of $f(x) = e^x$ to model the number of Facebook users? What parts of the graph are modeled more precisely with this function than with the functions previously explored (see figure 5.8)?

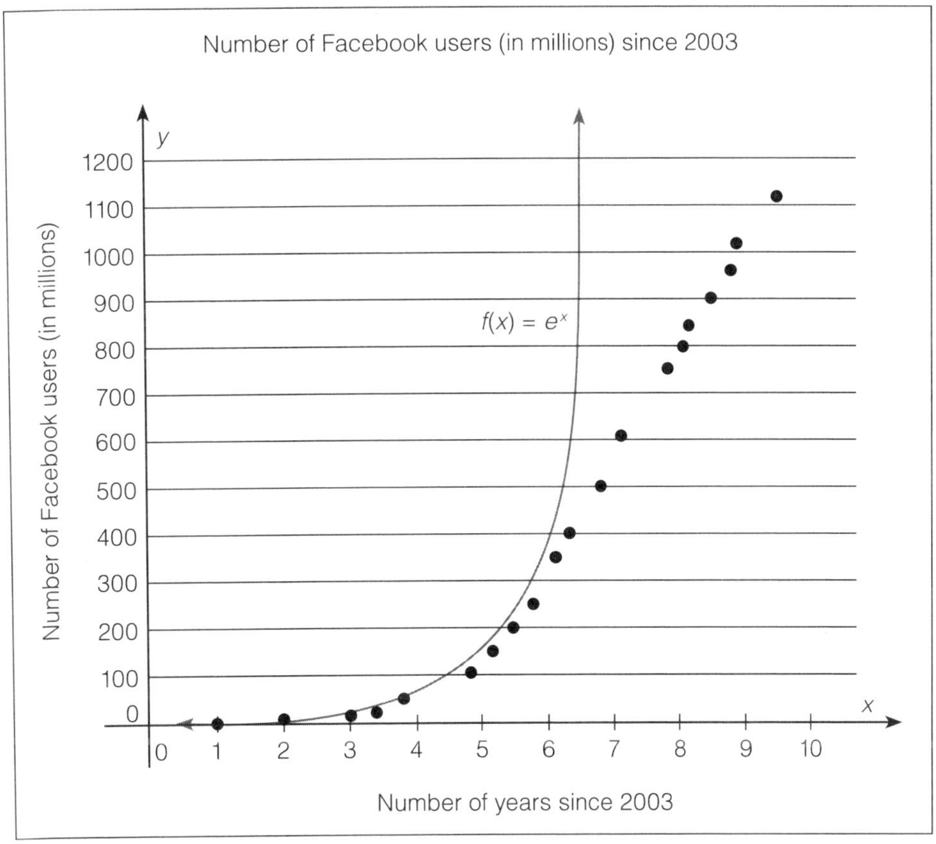

Figure 5.8: Exponential modeling of Facebook users.

It appears from the graph in figure 5.8 that the exponential model does provide a close fit for some pieces of the graph but needs to be adjusted before it can represent the entire data set. How would you revise the values of a, c, and k in the exponential function $j(x) = ae^{cx} + k$ to create the best fit to the data set? By strategically modifying the values of a, c, and k, you can create a model that aligns better to the data than the original function of $f(x) = e^x$. Compare the graphs of the two exponential functions provided in figure 5.9 (page 118).

What conclusions do you draw about the juxtaposition of the original graph of $f(x) = e^x$ with the graph of $h(x) = 0.8e^x + 5$? The adjusted graph $h(x)$ represents an improved model for the data set, as more data points are closer to the new graph than the original $g(x) = e^x$. Here is where you start to consider what is most important in your model. Is it more important for the graph to include as many points as possible or to indicate the general pattern of the data points? This needs to be guided by the context you are modeling. There will be some contexts where the accurate prediction is more important than the general shape

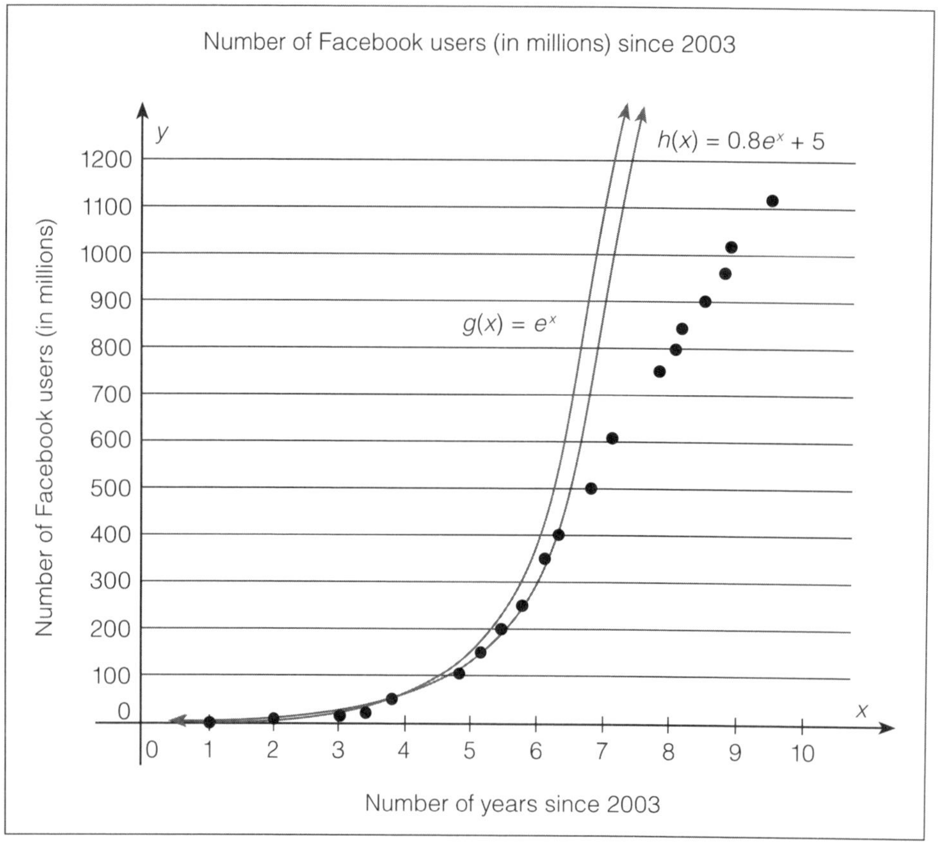

Figure 5.9: Modifying parameters in the use of exponential functions in modeling.

of the data, such as the difference in predicting the outcomes of medical treatments compared to the possible outcomes of sporting events. The role and the significance of understanding the impact of each parameter in different types of functions are important factors in your mathematical modeling. At this point, reflect on how effectively the quadratic model represents the number of Facebook users compared with the exponential model. A comparison is provided in figure 5.10.

What do you take away from the model comparison? Perhaps you notice that the exponential model addresses more points from $x = 1$ to $x = 6$. The quadratic model appears to be the better fit between the x-values of 6 and above. Is it possible to rely on *each* function type in order to explain, model, and make predictions throughout the entirety of the time interval? You may wonder about the last value in the data set. How would you describe a model for the number of users that includes the last data point as well?

How does the data point at 8.3 years after 2003 differ from the pattern in the years prior? As the function is an overestimate for this number of Facebook users, you may need a model that provides a different rate of change. What function types demonstrate a behavior that would suggest an increase in the function value and yet a decline in the rate of change of the number of Facebook users as suggested by the data? In other words, in representing the number of Facebook users from $x = 8.3$ and beyond, the model may need to represent a slowing in the rate of growth. What function types come to mind?

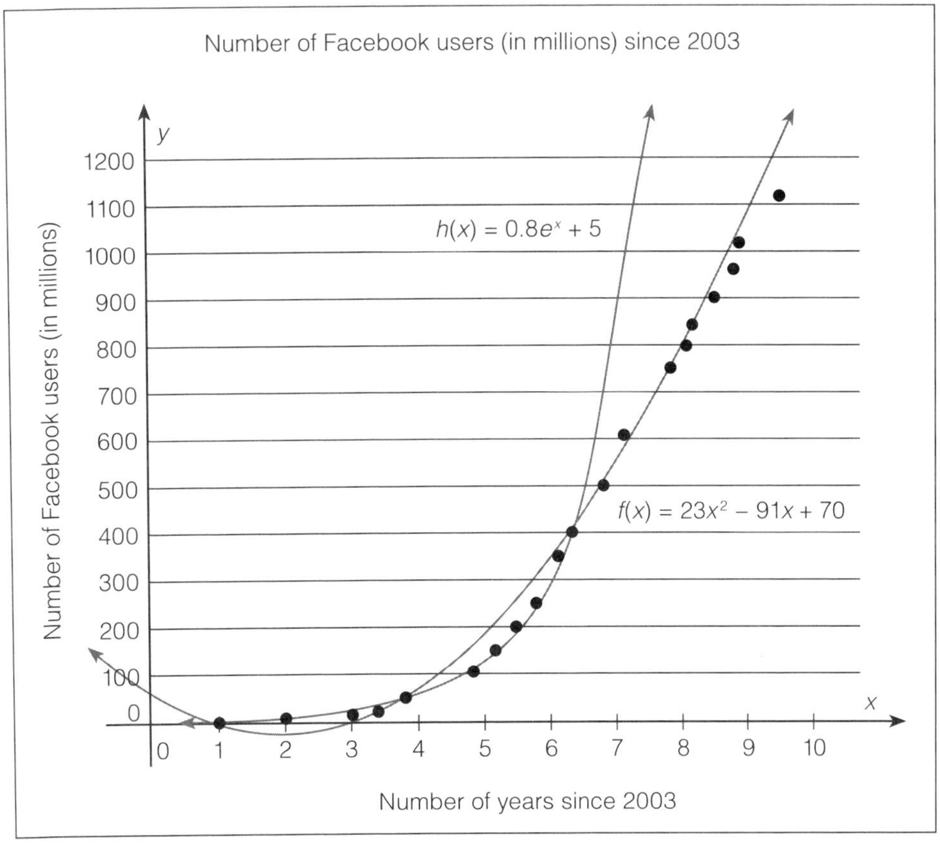

Figure 5.10: Comparison of quadratic model and exponential model.

Besides polynomial functions and exponential functions, students also investigate logarithmic and radical functions. The parent functions for the natural log function $f(x) = \log_b x$, or radical function, $f(x) = x^{m/n}$ (where m and n represent integers and do not have a common factor, with n not equal to 0; also, in order to represent a radical function, n should not equal 1) can illustrate a rate of change that is positive and yet a rate of growth that slows down as the x-values increase. Understanding characteristics such as the rate of change for multiple functions greatly aids in creating effective models for different contexts.

Modeling With Radical Functions

Revisiting radical functions to model the number of Facebook users—in particular for the x-values of 8.3 and beyond—may provide a good approach to represent and predict in this context. Similar to earlier conversations, modifying the parameters and the specific radical function to represent the given modeling task requires understanding how, when, and which parameters to adjust. How would you assign initial values to your parameters if you wanted to start your radical function model at the data point (8.3, 901)? What other parameters would you adjust from this starting point? To start at the data point (8.3, 901), you should start with the function $j(x) = a(x - 8.3)^{m/n} + 901$. This shifts the parent radical function to start at the given point. With additional adjustments to the parameters a, m, and n, you might obtain a graph similar to the graph in figure 5.11 (page 120).

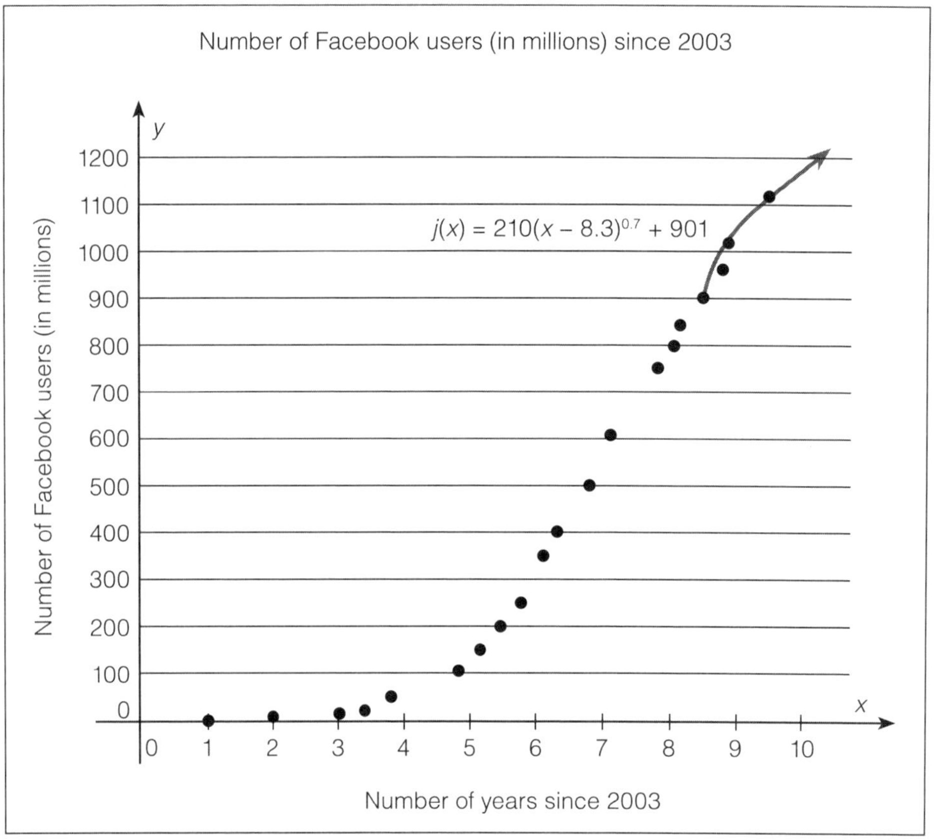

Figure 5.11: Modeling with radical functions.

Notice how the parent radical function has been changed to account for the portion of the graph beyond $x = 8.3$. The radical function is represented by $j(x) = 210(x - 8.3)^{0.7} + 901$. Because you shifted the radical function to align to a data point in the task, the domain of the model is the values of x at 8.3 and beyond. Also note that it is quite possible and in fact likely that you used a different combination of parameters to model this portion of the number of Facebook users.

Lastly, it seems that each of the three function types—quadratic, exponential, and radical (among others)—can model the number of Facebook users quite effectively but only for specific portions of the data. By looking at the three functions concurrently, you can observe the different portions of the graph and the manner in which they can work together to model the number of Facebook users for the entire time frame (see figure 5.12).

Modeling With Piecewise Functions

Take a moment to analyze which function should be used to model each specific portion of the graph related to time (x-values). Perhaps you notice that in the beginning, the rate of change for the first six years is well represented by the exponential curve. Therefore, the data suggest that an exponential model can describe the number of Facebook users starting in 2004 for about five years. Similarly from $x = 6$ to and including $x = 8.3$, the quadratic model can be used, followed by the exponential curve beyond $x = 8.3$. How do you express algebraic functions in pieces? Conveniently and purposefully, the term *piecewise function* refers to defining different functions to be used for specific portions of a domain. The

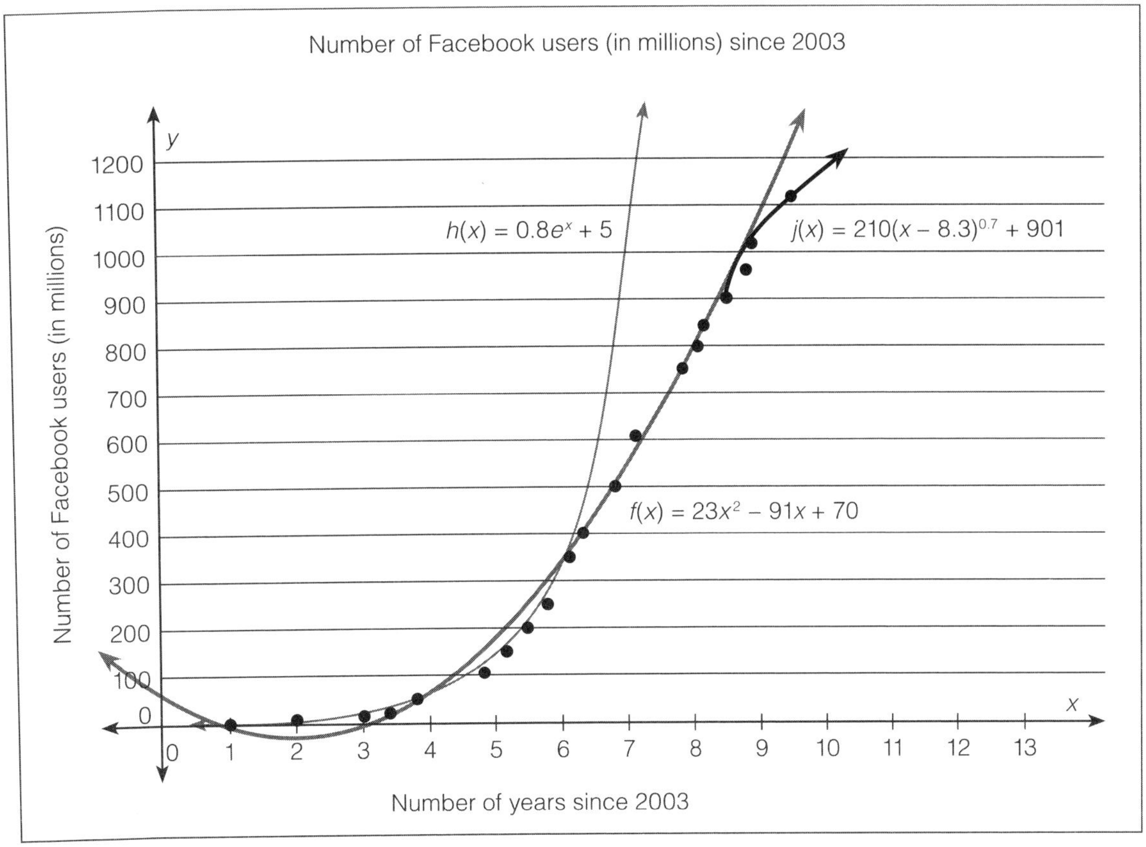

Figure 5.12: Modeling with various function types.

particular piecewise function that represents the number of Facebook users described here is provided in figure 5.13 (page 122).

Through analyzing the data points and reasoning with function types and graphs, you have explored one way to model with mathematics. By making sense of the context and providing a model that can be used to make predictions, many questions can be posed and subsequently answered, including:

- When would you predict the number of Facebook users to reach 1.5 billion? While you should always be cautious in making predictions that go beyond the domain or range of your data set, you could use your piecewise function to substitute in the value of 1,500 for *x* and determine an estimate (about 10.83 years after 2003, or around 2014). Note: Facebook achieved 1.5 billion users just after halfway through 2015 (Statista, 2015).
- How well would a linear model fit the data for the *x*-values between 5 and 7? What graphical relationship would you look for to support this claim? You see from the graph in figure 5.2 (page 110) and the data in table 5.1 (page 104) that a linear model might be an effective model for *x*-values between 4.8 and 7. Another way to help students see this relationship is to have them use uncooked spaghetti as a tool to lay on a drawn or printed graph to model where a linear relationship is evident.

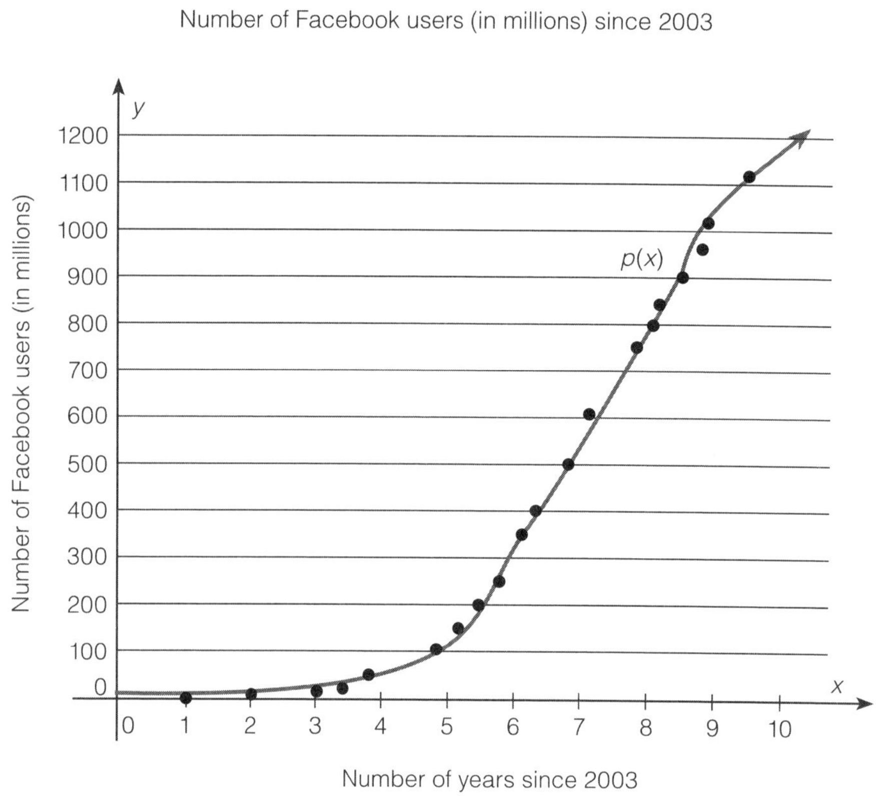

$$p(x) = \begin{cases} 0.8e^x + 5 & 0 \leq x \leq 6 \\ 23x^2 - 91x + 70 & 6 < x \leq 8.3 \\ 210(x - 8.3)^{0.7} = 901 & x > 8.3 \end{cases}$$

Figure 5.13: Modeling number of Facebook users using a piecewise function.

- How do the rates of change in specific time intervals compare with one another? You can see that the number of Facebook users gathers speed slowly at first then demonstrates a rapid rate of growth as illustrated by an exponential curve and finally tapers off in later years. What are the advantages and disadvantages of using a piecewise function to help in such an analysis? The obvious advantage of a piecewise function is that you can tailor different segments of the data set to different types of functions. The disadvantage of these functions is when you need to make a prediction beyond your defined data set, such as when there will be two billion Facebook users; it can be challenging to know what type of function will best model how the data will change.

The exploration of this task demonstrated how to use mathematical reasoning and sense making to better understand the number of Facebook users at different points in time. You were engaged in Mathematical Practice 4, "Model with mathematics," as you mathematized real-world contexts. You connected the number of Facebook users to different types of functions. Similarly, you analyzed and drew conclusions through the models that involved the use of tables, graphs, and quadratic, exponential, and radical functions. Lastly, you determined the quality of different models and made improvements.

In addition to Mathematical Practice 4, you also engaged in Mathematical Practice 2, "Reason abstractly and quantitatively." To understand, represent, and make predictions in this context, you had to abstract the relevant information and represent it symbolically while reasoning quantitatively with the number of years since 2003 (x-values) and the number of Facebook users (y-values). This decontextualizing and then recontextualizing is a focus of Mathematical Practice 2 as it entails the analysis and creation of a coherent way to represent the task while refining the understanding of the quantities. While symbolic manipulation and calculations are needed in this task, it was the mathematical reasoning and knowing how to draw upon the different function types that classified your engagement as Mathematical Practice 2. It must be noted that in order to engage effectively in the Mathematical Practices during instruction, the *learner* needs to be engaged in the thinking. As the teacher, you should support student focus on no more than two or three of these practices. It is not the absence of other practices, but rather the deliberate focus on a few, that supports a better understanding of each of the Mathematical Practices.

Modeling a Bird's Wings in Flight

Next, explore a different modeling task called *bird's wings in flight* in figure 5.14.

DO NOW

How would you model the bird's wings in flight as illustrated in the graph?

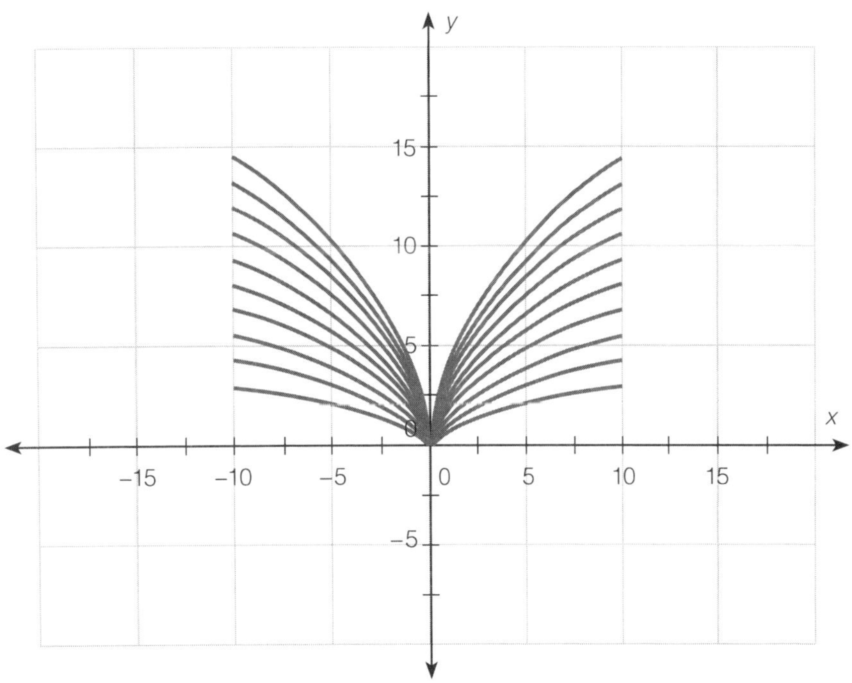

Figure 5.14: Modeling a bird's wings in flight task.

Mathematics enables you to represent various types of motion as often illustrated in animation, movies, art, and video games. In the bird's wings in flight task, the graph in figure 5.14 provides an approximation of what a bird's wings in flight might look like as a graph. This isn't *really* a bird in flight. One way that this is not real to life is that the wings appear to lengthen as they move from the bottom representation to the top one. However, it might be something that could be used to provide a digital animation to look like a bird's wings in flight.

How do you model the basic shape of the wings of the bird? There are a variety of function types to model and represent the wings of the bird. Perhaps you began by focusing on the wings in quadrant I. Did you consider a radical function such as $f(x) = \sqrt{x}$? The graph would provide the curvature of one of the representations of the wing, and if you restrict the domain to positive *x*-values, then you could represent the right wing of the bird. How would you create the wings of the bird in quadrant II? What mathematics content can you use in order to model the left half of the bird's wings?

In examining the graph of the bird's wings, you may notice that the right side and the left side are reflections of one another over the *y*-axis. How does your flexible understanding of mathematics provide the opportunity to consider different perspectives for this modeling task? Did you think about using a reflection about the *y*-axis to produce the left set of wings? Perhaps you took advantage of your understanding of the role of function parameters to make sense of this task. When you consider modeling with function types, you bring in your knowledge of geometry to support your thinking with algebraic and functional approaches.

How could you create additional representations of the wings of the bird? The first exploration involved looking at the square root function, which can be represented as $f(x) = x^{1/2}$. What other rational exponents could be considered within exponential functions to model the wings of a bird? What are the common characteristics of these types of functions (see the graph with the scale changed to highlight the differences in the types of functions in figure 5.15)?

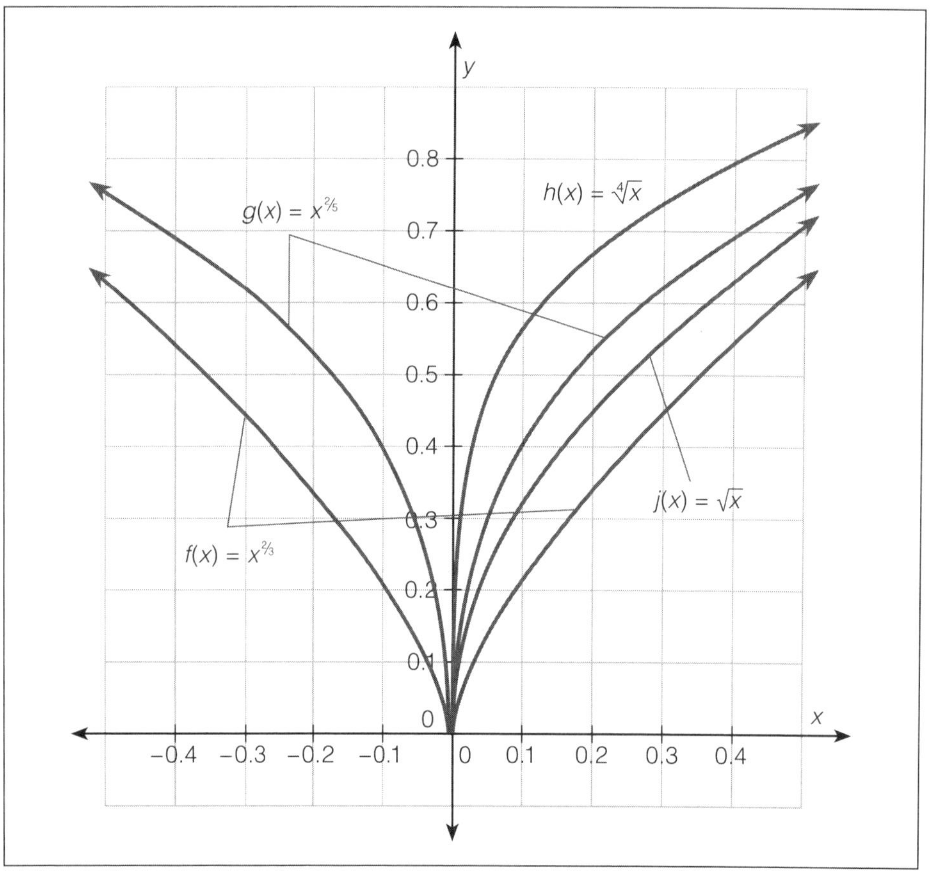

Figure 5.15: Exploring functions with rational exponents.

You may have noticed that different radical functions can provide different representations of the bird's wings. How does understanding the impact on the graph of changes in the exponent of a radical function provide an avenue for determining a model for this task? Having a flexible understanding of functions allows students to know *how* and *when* to use particular functions as well as to know how to adjust functions to create better models or different representations. It is this type of engagement with the content that makes modeling a rich and engaging opportunity to explore the use of functions throughout high school mathematics.

Complete the task to think about what functions you could consider as models for the wings of the bird (see figure 5.16)?

DO NOW

Consider this list of possible functions to model the bird's wings in flight.

1. $y = \sqrt{x}$ if $x \geq 0$ and $y = \sqrt{-x}$ if $x < 0$
2. $y = \sqrt[4]{x}$ if $x \geq 0$ and $y = \sqrt[4]{-x}$ if $x < 0$
3. $y = x^{2/3}$, for all values of x
4. $y = x^{2/5}$, for all values of x
5. $y = ln(x + 1)$ for $x \geq 0$ and $y = ln(-x + 1)$ for $x < 0$
6. $y = -e^{-x} + 1$ for $x \geq 0$, and $y = -e^{x} + 1$ for $x < 0$

How would you determine the appropriateness of each of these potential functions as effective models for the wings of the bird?

Figure 5.16: Function types and related parameters and constraints.

Consider numbers 1 and 2 in figure 5.16 (see figure 5.17).

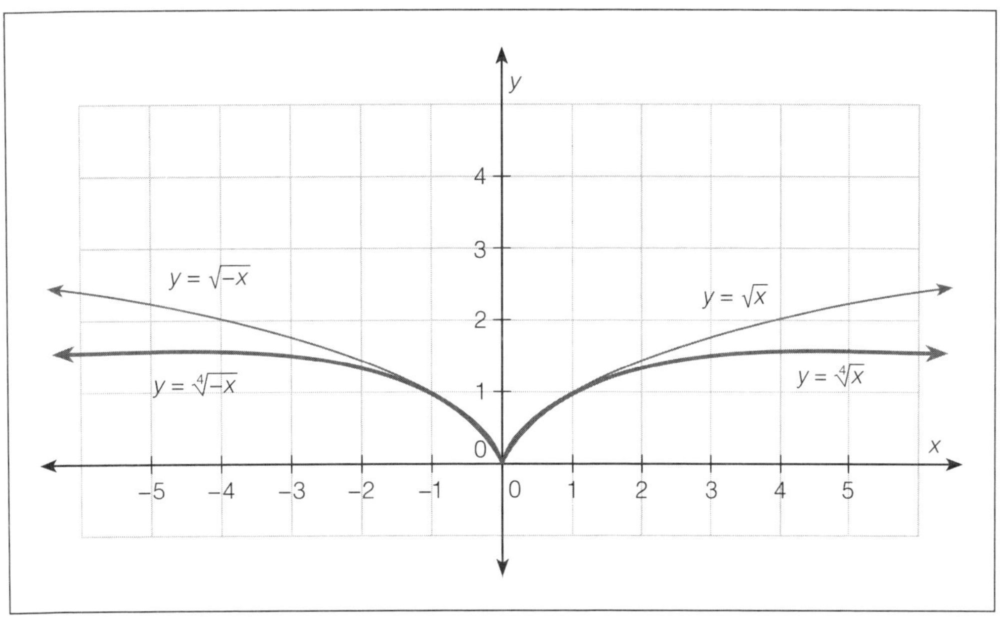

Figure 5.17: Graphs of $y = \sqrt{x}$ for $x \geq 0$, $y = \sqrt{-x}$ for $x < 0$, $y = \sqrt[4]{x}$ for $x \geq 0$, and $y = \sqrt[4]{-x}$ for $x < 0$.

In each of these options, there are two separate functions represented. Why would you need two separate functions to represent the bird's wings? What is the impact of having $(-x)$ under the square root for x values that are smaller than 0? You notice that due to the restrictions on the possible values of these radical expressions, the type of x-values would be restricted. Having two different functions defined in this manner produces the wing shape on both sides of the y-axis. By having both $y = \sqrt{x}$ if $x \geq 0$ and $y = \sqrt{-x}$ if $x < 0$, you model one pair of wings.

In the case of numbers 3 and 4 in figure 5.16 (page 125), note the absence of the two pieces for positive and negative x-values (see figure 5.18).

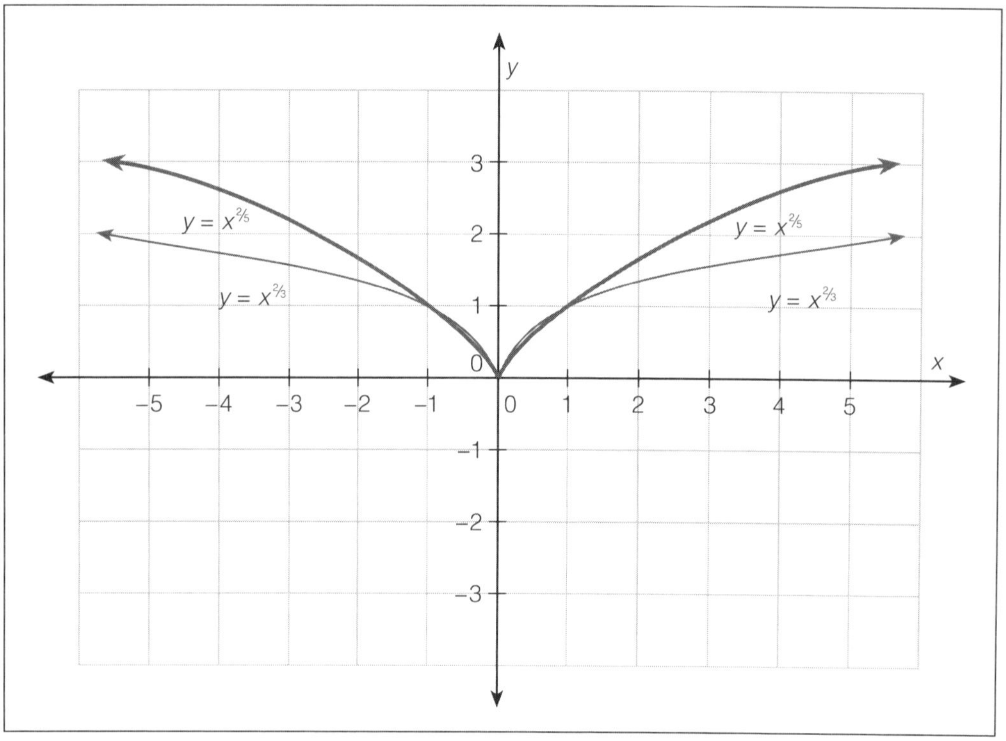

Figure 5.18: Graphs of $y = x^{2/3}$ and $y = x^{2/5}$.

What common characteristics do $y = x^{2/3}$ and $y = x^{2/5}$ (numbers 3 and 4 in figure 5.16) possess? What role does the denominator of these rational exponents play in determining the domain of these radical functions? Recall that for the square root and fourth root examples in numbers 1 and 2, there was a domain restriction due to the mathematical structure of these functions. Here, the odd denominator of the exponent—the index of the radical—creates a function that can accept both positive and negative input values. Why are the numerators in both exponents even? Notice that $y = x^{2/3} = (x^{1/3})^2$. As a result, while the cube root expression inside the parentheses has both positive and negative outcomes, the given expression will have a range of strictly non-negative values. The same is true for the fifth root function. A modeling problem such as this one has the potential to reinforce the simultaneous understanding of function types, their graphs, rational numbers, and exponents.

The next two candidates for modeling the wings of the bird include the natural logarithmic and exponential functions (see numbers 5 and 6 in figure 5.16, page 125). Explore how each possesses the curvature needed to represent the wings of the bird (see figure 5.19).

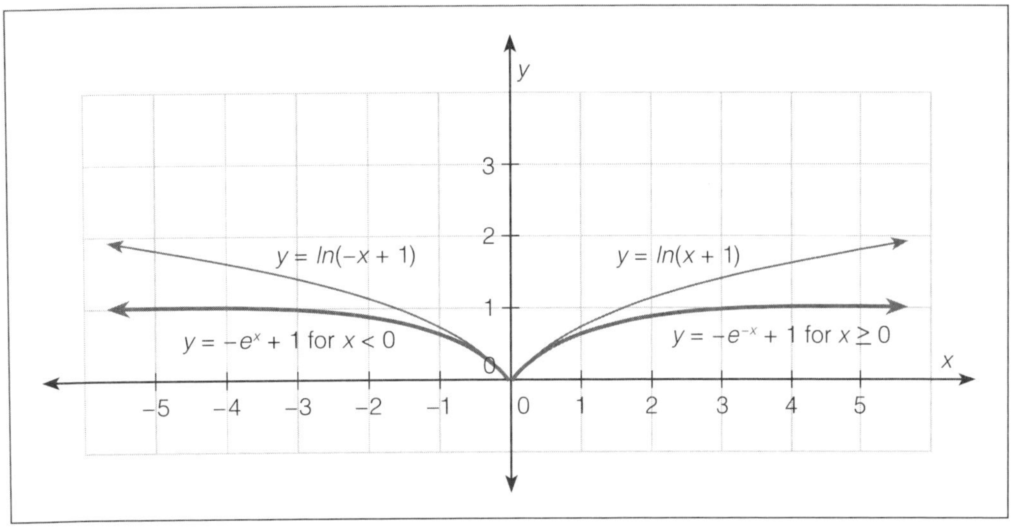

Figure 5.19: Graphs of $y = ln(x + 1)$ for $x \geq 0$, $y = ln(-x + 1)$ for $x < 0$, $y = -e^{-x} + 1$ for $x \geq 0$, and $y = -e^{x} + 1$ for $x < 0$.

Perhaps you noticed that by performing a horizontal shift of one unit and restricting the domain of the parent natural logarithm function, the desired shape and placement are achieved. What is the connection between a horizontal shift of one unit for the natural log function and a vertical shift of one unit for the exponential function? Why do these two strategies both demonstrate a shift of one unit, even though the *x*- and *y*-orientation of the shift has been inverted? The connection here demonstrates the inverse relationship between the exponential and logarithmic functions explored in chapter 4. The relationship between a function and its inverse interchanges the dependent and independent variables so that the horizontal shift in the natural log function becomes a vertical shift in the inverse exponential function. Understanding the relationship between functions allows for understanding how transformations will impact each function differently.

At this point, you may recall that the wings of the bird need to have limited domains in order to correctly model the representation shown in figure 5.14 (page 123). How can you constrain or limit the wings of the bird? You can see from the original graph that the *x*-values take on values between –10 and 10. As a result, one approach would be to adjust the model to limit the domain to $-10 \leq x \leq 10$. In this manner, not only will the functions provide the shape desired, but by limiting the domain, you are limiting how far out the bird's wings will reach. This allows the precision needed to model the flight of the bird accurately, representing Mathematical Practice 6, "Attend to precision." The next section of this chapter, The Classroom, will further describe how students engage in connecting representations and modeling with mathematics.

The Classroom

To this point, you have explored modeling using different functions and strategies. Now, you will have the opportunity to explore what this may look like in a classroom setting. The following videos provide perspective and valuable insights into students' conceptions and misconceptions regarding functions and mathematical modeling.

Modeling with mathematics involves mathematical reasoning, as witnessed by both the Facebook users task and the bird's wings in flight task. Mathematical reasoning with functions begins in the middle grades as students explore connections between different representations of functions. In the first video, students in grade 8 are engaged with a task that connects tables, graphs, and equations. You are strongly encouraged to watch the video in its entirety prior to proceeding.

www.solution-tree.com/Connecting_Representations_of_Functions

Now that you have explored the video, focus on the manner in which students are engaged in making sense of the task. How do students in the middle grades begin to reason with multiple representations such as tables, equations, and graphs? What different strategies are evident? Often, it is easier for the teacher to provide one strategy and ask students to apply this strategy. For example, in this case, the teacher could have said, "Start with the equation, plug in points in the table, and plot them in the graph." This would take less time; however, it would remove the role of the student to actively think about and engage in the problem. Instead, by giving the task to sort the representations, the teacher helps students develop multiple strategies to determine how the tables, graphs, and equations are connected. This provides a richer foundation and a deeper understanding of functional relationships and representations. Some students begin with the table and test the points in the equations. Others begin with the graph and see if the points in the table are represented on the graph. Still others begin with the equation and make sense of the table and graph based on each equation. Students need experiences that allow them to connect and reason with multiple representations and then build on this foundational understanding as they gain knowledge and familiarity with various function types in high school.

In the video, how does the teacher facilitate the discussion through questioning? Perhaps you noticed that the teacher does not provide explicit answers that indicate whether the student is correct or not, but rather asks questions to prompt student thinking. The teacher purposefully guides the conversation toward the learning goal of connecting multiple representations and the type of information students can gather from each representation. The teacher guides the lesson by asking questions such as "What's an advantage of using a graph?" and "What's the advantage of using an equation?" Upon gathering student responses, the teacher follows up by posing the question, "What is an advantage of using a table of values?" He connects to the big idea of the lesson by asking the critical question, "Why do we have all three representations?" The student states, "They are tied together visually and numerically." When students have such experiences in the middle grades reasoning with equations and multiple representations, high school experiences can continue to develop their mathematical understanding with different families of functions.

In this video, the students engage in Mathematical Practice 3, "Construct viable arguments and critique the reasoning of others." As students work in small groups, they support their own reasoning and listen to the strategies of other students in their group. They then compare their strategies with the strategies of other students in the class.

In the second video, you will explore how students engage in modeling the bird's wings in flight task. Recall The Classroom section in chapter 4 related to how students explore different types of functions. Knowing how students reason with and analyze function types is critical in understanding how they approach modeling tasks. In this next video, students engage in the task described in figure 5.14 (page 123). You are strongly encouraged to watch the video in its entirety prior to proceeding.

www.solution-tree.com/Modeling_Bird_Wings_in_Flight

How are students modeling the wings of the bird using their understanding of functions? Upon observing what students are attending to in their small groups, the teacher asks one of the students working in a small group, "Can you show me how you got the graph that you have?" The student responds with, "I used the square root of x, square root of $-x$, and kept increasing the number of xs in each one." In this short exchange, you gain a sense of the type of function that the student is using to model the wings of the bird. When asked by the teacher, "The number of xs . . . What do you mean by that?" she answers, "Two x, three x, four x, . . . to make the bird go up." The student then proceeds to show the teacher her graph (see figure 5.20, page 130).

Notice that in order to make sense of the bird's wings in flight, the student is changing the coefficient of x to produce the multiple instances of the bird's wings. She describes ways that she reasoned repeatedly with how changing the value of the coefficient of x from 2 to 3 to 4 would change the bird's wings. Similarly, she engaged in the same process for the other half of her model and used similar reasoning strategies to modify her equations from $y = \sqrt{x}$, $y = \sqrt{2x}$, and $y = \sqrt{3x}$ to $y = \sqrt{-x}$, $y = \sqrt{-2x}$, and $y = \sqrt{-3x}$. To be successful in modeling with mathematics and continue to make sense of mathematizing the world, students need to engage in Mathematical Practice 8, "Look for and express regularity in repeated reasoning."

Student modeling the bird's wings in flight task using the square root function:

$y = \sqrt{x}$, $y = \sqrt{2x}$, $y = \sqrt{3x}$ as well as $y = \sqrt{-x}$, $y = \sqrt{-2x}$, $y = \sqrt{-3x}$

Figure 5.20: Student modeling bird's wings in flight.

TQE Process

At this point, it may be helpful to watch the second video (page 129) again and pay close attention to the task, questioning, and opportunities to collect evidence of student learning.

The TQE process can help you frame your observations. Teachers who have a deep understanding of the mathematics they teach:

- Select appropriate *tasks* to support identified learning goals
- Facilitate productive *questioning* during instruction to engage students in Mathematical Practices
- Collect and use student *evidence* in the formative assessment process during instruction

The *task* chosen for this lesson involves the mathematical modeling of a bird's wings in flight. The learning goal is for students to see that there are various ways to model the wings of the bird mathematically using their understanding of different function types. The task supports this learning goal by allowing students opportunities to explore function types—such as square root functions, other radical functions, and even exponential and logarithmic functions. Students draw on the mathematics they have access to and use their knowledge of functions and parameters to manipulate functions of their choice in an effort to model the wings of the bird. During the lesson, students engage in a productive struggle that causes them to revisit and continually modify their initial approach to model the task accurately. This aspect of the task supports a critical element of the learning goal.

The *questioning* the teacher uses engages students in meaningful discussion and allows for the opportunity to explore what the students are thinking while in their small groups. The teacher asks the students to explore the task and then engages one of the small groups by asking them questions regarding which type of function they are using and whether that function explains "the whole graph." Since the wings of the bird are reflected about the *y*-axis, the teacher begins to look for ways in which the students reason with functions that demonstrate *y*-axis symmetry. One student mentions, "The first thing I noticed was an absolute value function." The teacher asks a question focused on the reasoning behind this selection to gather further evidence of student learning. In order to get at mathematical precision regarding the graph of absolute value functions, the teacher poses the question, "What does an absolute value function's graph do?" The same student motions that the picture of the bird is "kind of like a V shape . . . it's kind of reflected."

The teacher uses *evidence* of student thinking and understanding to help him navigate the discussion through this task. In order to gain a better sense of students' thinking, the teacher pushes for further elaboration by asking, "How exactly does it reflect the function?" to which a second student responds, "Whatever negative numbers are on the left side of the graph would be positive numbers on the right side of the graph." Notice that by engaging the students in this discussion, the teacher gains some insight into *what* and *how* each student is thinking about the reflection, deepening the teacher's perspective and providing valuable insights into student thinking. The first student seems to be looking at the task functionally (absolute value) and graphically, and the second student reasons about the reflection by describing it numerically. Such engagement opportunities are valuable to the students and the teacher in gauging the progress of student understanding related to the specified mathematical learning goals.

The design of the task involved the simulation of the wings of the bird in flight and mathematically producing the images that would simulate the different positions of the wings when the bird flaps its wings. Here, the students were asked to use their knowledge of mathematical concepts such as functions and their graphs to make sense of, represent, and produce a desired result. As such, the students were actively engaged in Mathematical Practice 4, "Model with mathematics." Additionally, in modeling with mathematics, a great deal of precision is necessary to achieve the desired results that match the context of the modeling task. It is critical for teachers to engage students in Mathematical Practice 6, "Attend to precision," to ensure clarity in their use of mathematical notations and definitions, such as the exchange in regard to the negative square root function and differentiating expressions, equations, and functions. It is important to emphasize the clarity of communication in being precise in the language used by the students.

The Response

In modeling tasks, students sometimes struggle with how to use the mathematics they know to represent and analyze given situations. There are times when there is a disconnect between different types of understanding of mathematical topics—such as understanding the connections between different representations of functions and the ability to apply those understandings meaningfully to model real-world and mathematical situations. For instance, in the Facebook users example, students may create a model that is only appropriate for one part of the graph. In thinking about the bird's wings in flight task, they might model a bird with infinitely long wings unless they attend to restricting the domain of the selected function. The flexible and accurate use of different types of functions comes with experience with these types of tasks. This is more likely to occur when such opportunities are built into the lesson with the expectation that models can often be improved. Optimally, you would like students to be comfortable using multiple strategies and different functions to model real-world events. In modeling contexts, not all functions that are appropriate representations may be applicable due to physical, financial, or time constraints. Therefore, by possessing a flexible and coherent understanding of mathematics, students can engage effectively in modeling tasks.

Teachers can provide these experiences by comparing and connecting function types early and often throughout high school mathematics experiences. The relationship between the quadratic and square root functions and the exponential and logarithmic functions are such connections. By having students investigate tasks that explore the behavior and characteristics of functions, teachers can provide opportunities for students to reason and conjecture about appropriate uses of functions as tools in modeling tasks. Modeling with mathematics provides a wonderful chance to make sense of, analyze, problem solve, and in essence, mathematize the world. When students explore a problem in context and think through the mathematics to be able to predict or model a given situation, they illustrate a connected sense of the mathematics.

Reflections

1. What do you feel are the key points in this chapter?
2. What challenges might you face when implementing the key ideas related to modeling with mathematics? How will you overcome them?
3. What are the important features for developing an understanding to use mathematics to model different contexts, and how will you ensure your instruction embeds the support needed for these features?
4. Select a recent lesson you have taught or observed focused on modeling. Relate this lesson to the TQE process.
5. What changes will you make to your planning and instruction based on what you read and considered from this chapter?

CHAPTER 6

Statistics and Probability

This chapter connects statistics and probability concepts to the reasoning and sense making you have explored from algebra and geometry. Some tasks will connect with the reasoning and justifying aspects of the previous chapters, such as reasoning how to make predictions using information from a real-world situation, justifying decisions made from the analysis of a given data set, or examining how sample surveys, experiments, and observational studies can be used as mathematical models. Other tasks will look at how the analysis of a data set connects to the shape of the data display and measures how well a function models the pattern of a data set.

By definition, *statistics* is the collection, analysis, and interpretation of data. *Probability* is the ratio of successful outcomes to the total number of outcomes. These two topics, along with the use of randomness, are important concepts in mathematical modeling and simulating real-world events.

The Challenge

One important aspect of making sense of statistics and probability at the high school level is understanding the power of inference—the ability to reason and support the making of predictions from data. Inference connects to using functions to model patterns in a data set and making predictions based on those models. Concepts from statistics and probability are important to support decisions that you make in real-world situations. How are you preparing students to be critical consumers of the data they encounter on a daily basis? Consider the claim in figure 6.1.

DO NOW

A start-up company states that its new vitamin causes weight loss. It has asked you to invest in the company. What questions would you ask to determine whether its claim is true?

Figure 6.1: Making claims task.

It is likely that you would want to ask about the data used to support this claim. You might also want to know how the data were collected.

Imagine the company stated that it knows that its vitamin causes weight loss because it conducted a study during which it gave the vitamin to two hundred people and the overall average weight loss was over five pounds per person. Would this be enough information for you to invest in the product? It is likely that this information would lead you to other questions. What else do you need to know in order to feel confident that its claim is valid? You probably want to know how the study was conducted, such as whether these participants were randomly selected or if there was a characteristic that the participants all shared. Why is this information about how the study was conducted important? Knowing the method of the study is an important factor in assigning value to the findings from the data. Maybe many of the participants recently joined a gym, which could be an influencing factor in the weight loss. Were different factors such as joining a gym controlled for in the study? You may also be interested in how much weight was lost by each participant, in addition to the overall average. Did each person lose about the same amount, or were there some who lost a great deal compared to many who lost very little or even gained weight? If you learned that twenty participants lost sixty pounds each and the remaining 180 participants gained an average of one pound, are you still interested in investing? To make wise choices, you need to know how the participants were selected, how the design of the study accounted for all variables, the variation within the data set, and how the data were analyzed and presented. The appropriate design of studies, surveys, and experiments is included in a later section of this chapter.

This study examines the relationship between the use of a new vitamin and weight loss. You need to consider whether the difference cited in the weight loss is actually *caused* by the new vitamin or simply *correlated* with taking the new vitamin. A correlation occurs when two events tend to vary together. For example, think about the relationship between buying an ice cream cone and getting a sunburn. If you were to collect data from people over a weekend, you might find that more people who buy ice cream cones get sunburns than those who do not buy ice cream cones. These two events appear to be related. However, it is probably clear to you that buying an ice cream cone does not cause one to get a sunburn. This relation is more likely explained by the weather. On hot, sunny weekends, people are more likely to buy ice cream cones *and* they are more likely to get sunburns. These two events are *correlated* because they occur together but one does not *cause* the other. That is an important distinction between correlation and causation. In order to interpret results appropriately, you need to know if the design of a study warrants a conclusion of causation versus simply showing a correlation. The design of a study is a key factor in being able to properly claim the effect of one variable upon another. It is important to distinguish when studies and experiments present a correlation and when they demonstrate causation. A correlation would likely not provide a compelling case for investing in the start-up company described in figure 6.1 (page 133), but if the study warranted the designation of a causal relationship between the new vitamin and weight loss, an investment in the company might be a prudent decision.

It is important in the study of statistics and probability to base conclusions on appropriate evidence. The concepts explored in statistics and probability also include features about data sets, data displays, and rules of probability that help in understanding the process of making decisions and predictions about the world. These concepts will be explored throughout the chapter.

The Progression

In high school, students build on their statistical reasoning experiences from the middle grades. The progression of statistics and probability concepts includes analyzing single variable and bivariate situations, developing rules of probability, and applying the rules and reasoning of probability to model the world. Students analyze data sets, model situations, and make inferences and predictions of actual real-world events. Underlying these concepts is the way in which randomness can have a role in predicting future events. The understanding of randomness and the differences between making predictions for individual and group outcomes has an important impact on making sense of patterns within data, including how functions can be used to help make predictions from data. The following is a progression of the development of these concepts.

- Create and interpret simulations, sample surveys, experiments, and observational studies.
- Describe and compare data sets.
- Use characteristics of data to make predictions (inference).
- Describe, apply, and analyze normal models.
- Determine when causal relationships can be implied.
- Fit functional models to bivariate data and interpret characteristics of these functions in context.
- Examine frequencies and relative frequencies in two-way data tables.

Middle School

In middle school, students analyze, display, and describe data sets by shape, center, and spread. They make inferences about the differences between given data sets. Students develop an understanding of probability, simulate real-world events, and learn how randomness plays an important role in determining the validity of an inference. Students create and interpret scatterplots and examine the relationship between the variables represented on a graph, looking at lines that fit the data in an informal way.

High School

In high school, students build on understandings from middle school to describe data sets by their shape (uniform, bimodal, normal, and so on), center (mean and median), and spread (mean absolute deviation, standard deviation, and range). They consider the impact of how the data are collected on the interpretation of the data. Students also determine when and how to make inferences and predictions and the degree of accuracy they can assume about those inferences and predictions.

Students consider appropriate sampling techniques and the likelihood of particular outcomes from a given sampled set of data. When data sets model the normal distribution, analysis leads to inferring relationships about the larger population due to the characteristics of this shape of the data. Students explain how sample surveys, experiments, and observational studies can be used to provide information about the world around them. They learn to both create and interpret these statistical investigations. Students examine whether factors have been accounted for in research designs in order to allow for generalization to a larger population. They also determine the margin of error that needs to be considered for a generalization.

Students build on a basic understanding of probability in order to support the development of the difference between individual and group outcomes. While the prediction of a group of outcomes—such as determining the number of people in a population who are likely to purchase a new car within the next year—can be made with some degree of confidence, it is more difficult to predict individual outcomes, such as the likelihood of a specific individual purchasing a car within the next year. Students explore the difference in the probability of these outcomes through examining the results of data collection models, simulations, and two-way tables.

Bivariate data are represented using scatterplots. Students consider how patterns in the data lead to the use of different types of functions to make predictions based on the data. Students use mathematical models to interpret the relationships of real-world events. The degree to which functions model a given set of data and context is measured and analyzed. The relationship between correlation and causation is explored to determine how to provide appropriate context for predictions. As the focus of this chapter involves the connection among statistics, probability, and topics discussed in previous chapters, not all of the concepts within statistics and probability can be discussed here. For example, the role of independence in probability and the development of the rules of probability are not emphasized in this chapter.

The Mathematics

This chapter will explore connections between statistics and probability as well as how these topics connect to algebra and geometry. High school students must be able to apply the normal model; compare data sets; understand sample surveys, experiments, and observational studies; investigate probability; and explore scatterplots, both in the context of real-world situations and through mathematical models.

Applying the Normal Model

What does it mean for data to follow a normal pattern? How do you know when data can be analyzed using the characteristics of a normal distribution? One way to represent data is to create a histogram. *Histograms* show the frequency of the elements of a data set organized into intervals. This allows you to examine how the data are clustered across evenly spaced intervals. When you create a histogram of a data set, if the general shape of the bars follows the curve of a bell, then you have a normal distribution for the data. The normal model is an important representation and appears in many different data sets. It allows you to make inferences about the population to which the data apply. The general curve will create a peak in the middle of the histogram (near the important measure of central tendency of the data set, which is usually the mean) with decreasing frequency bar heights as you move away from the peak of the histogram. When the data follow a normal distribution pattern, you have certain tools available to analyze the data. How do these tools work in practice (see figure 6.2)?

How did you determine how often you should expect a weight loss of more than twenty-five pounds in the overall population? How confident are you in the answer you determined? Do you *know* that you are right, or do you *think* that you are right? This is one of the challenges in statistics—your answer is based on only a sample of the entire population. The only way to truly know that you are right is to examine the weight change for every member of the entire population. However, it is difficult to learn such information from every member of a population, so you need to be able to analyze the characteristics of an entire population based on information from a subset of the population. This means that you will have some uncertainty in your answer, but you will be able to infer an answer based on an analysis of the sample data.

How can the shape of the data set help you in your analysis? You see in the histogram that there is a single peak in the middle of the data set (in the bar that includes the value of the mean) and decreases on both sides of that peak. This indicates that the sample is normally distributed. What does this imply for the entire population? The key is whether or not this is a representative sample of the population. If this

DO NOW

Consider the weight loss study from figure 6.1 (page 133). A hypothetical data set from the study is provided in the appendix (page 159). The overall mean for this data set is a loss of 5.02 pounds (represented as –5.02); the median is a loss of 5 pounds; and the standard deviation is approximately 9.87 pounds. The histogram of the data set is shown here.

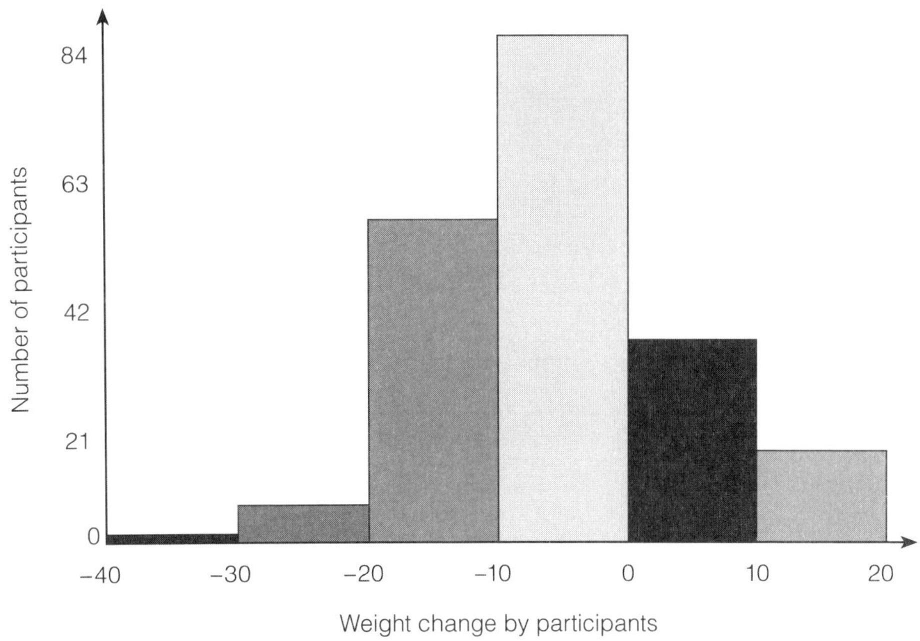

Based on the sample, how rare would a weight loss of more than 25 pounds be for any person who may take the vitamin in the future?

Figure 6.2: Analyzing weight loss.

is a representative sample, you can infer that the overall population will follow a similar pattern as the representative sample. This is an important fact to remember—you are always basing your analysis on the fact that the entire population will be normally distributed because the representative sample follows that pattern. Sometimes, you might be wrong. However, you base your analysis of the population on the sample that you selected, and if that sample is representative of the population, your inference is justified. The fact that the distribution of the weight loss follows a normal distribution pattern provides you a way to estimate the number of people who may lose a given amount.

How does knowing that the data set is normally distributed help you? You can use the fact that the data set is normally distributed to predict approximately how often a weight loss of more than twenty-five pounds occurs in this population. First, when a data set is normally distributed, you use the mean as the middle point of the data set, or the 50th percentile. The mean of this data set is -5.02 pounds (a loss) and the standard deviation is about 9.87 pounds. Standard deviation is a measure of the spread of the data. In a normally distributed data set, approximately 68 percent of the data set is within one standard deviation of the mean, approximately 95 percent of the data set is within two standard deviations of the mean, and approximately 99.7 percent of the data set is within three standard deviations of the mean (see figure 6.3).

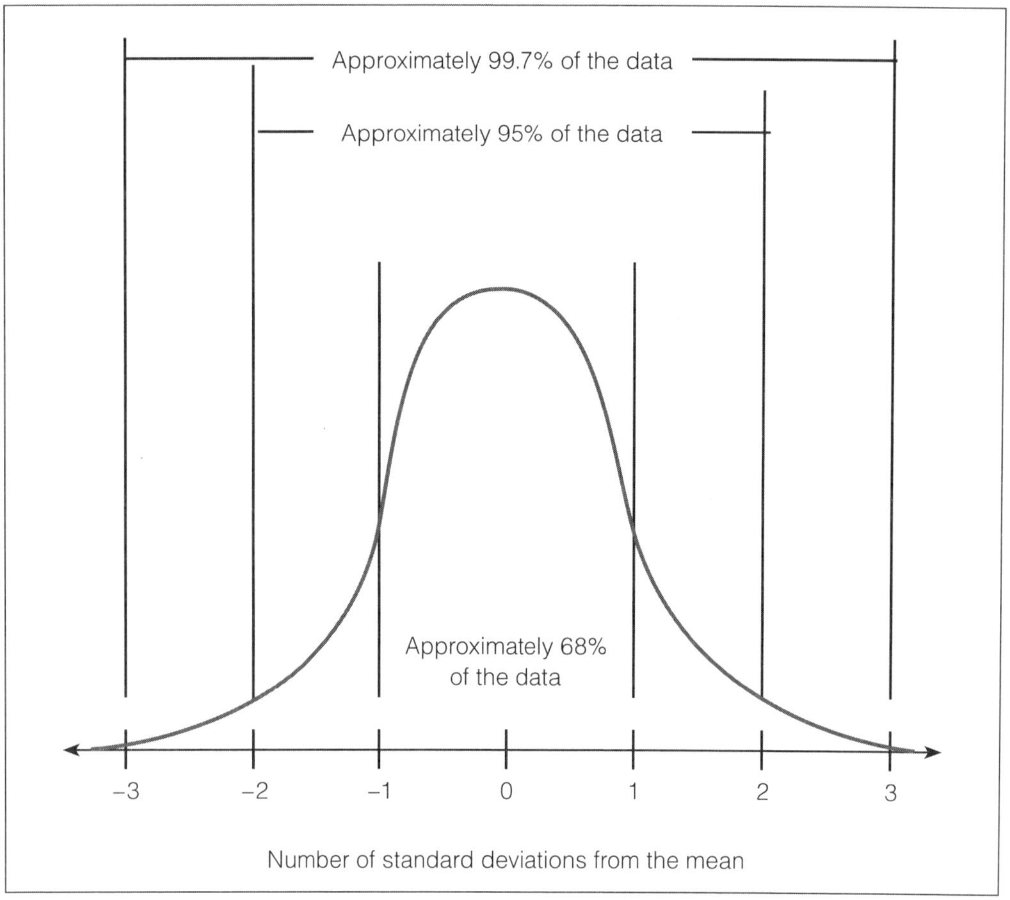

Figure 6.3: Spread of the normal distribution.

How does this information help you answer the question in figure 6.2 (page 137)? With a mean of −5.02 (a loss of a little more than 5 pounds) and a standard deviation of about 9.87, you know that 68 percent of the data should be between a weight loss of 14.89 pounds and a gain of 4.85 pounds; 95 percent of the data should be between a loss of 24.76 pounds and a gain of 14.72 pounds; and 99.7 percent of the data should be between a loss of 34.63 pounds and a gain of 24.59 pounds. How does this help you answer the question? Since the data are distributed on both the sides of the mean, as well as outside each of the standard deviation ranges, the additional 5 percent outside of the 95 percent range should be split evenly on each side of the mean. This indicates that about 2.5 percent of the data should be more than a loss of 24.76 pounds and about 2.5 percent of the data should be above a gain of 14.72 pounds. This means that you would expect about one in forty people to have more than a 25-pound weight loss.

The inference you developed in determining the likelihood of a weight loss of more than 25 pounds is based on the normal distribution. Normal distributions allow you to make inferences about the frequency of ranges of values, as well as specific values, in a data set. It is important to remember that the analysis you completed in this task is only valid when you are confident that the normal distribution model can represent the data set. There are times when a data set does not follow the patterns of the normal model, meaning that the data set is not clustered around the mean with decreasing frequencies as you move away from the mean. Considering and using the structure of the data set to analyze the data correctly are similar to considering the correct type of function you use to model a real-world situation. The structure of the normal distribution model allows you to make predictions based on data. In doing so, you are engaging in Mathematical Practice 7, "Look for and make use of structure."

Comparing Data Sets

Another important aspect of statistical reasoning at both the middle grades and high school involves comparing data sets. Being able to distinguish important characteristics between data sets is similar to distinguishing between linear, quadratic, and exponential functions. The sophistication of the data analysis distinguishes this experience in high school compared to previous explorations in the middle grades. What strategies can you use to compare data sets? What strategy do you think is appropriate to address the task in figure 6.4?

DO NOW

In the weight loss study, 100 of the participants were male and 100 of the participants were female. Examine the two histograms here.

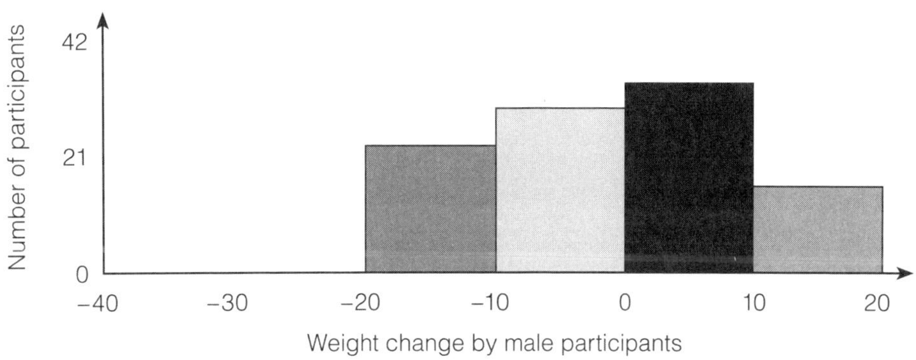

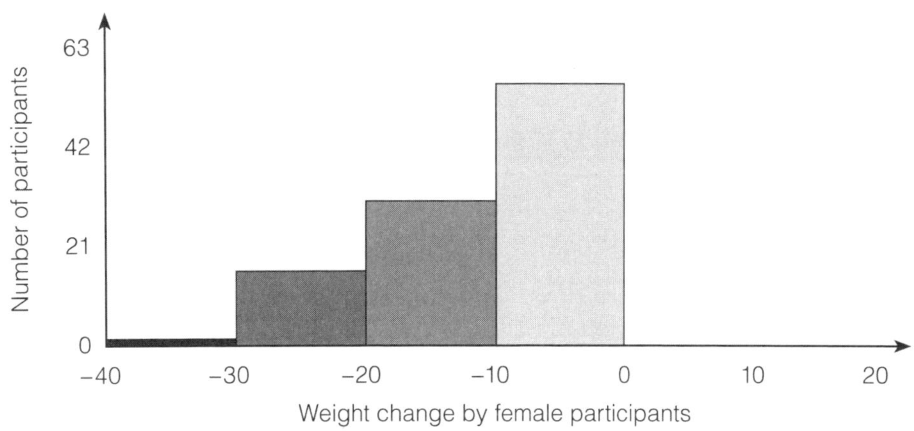

How would you describe the differences between the two data sets?

Figure 6.4: Comparing data sets task.

What do you notice about the two histograms? The histogram for the male participants represents a somewhat balanced data set, with a peak in the middle and the frequency of data elements decreasing on both sides, though not in the same decreasing pattern as the overall data set shown in figure 6.2 (page 137). However, the histogram representing the change in weight for the female participants demonstrates a skewed relationship, or a representation where the data are not balanced on both sides of the mean. The data are clustered to one side of the histogram. For the data representing the weight change for female participants, the histogram demonstrates a left skew, or a negative skew, as the data are clustered on the right side of the plot with a "tail" to the left, or toward negative values. The term *negative skew* is based on the direction of the tail, not a reference to the negative values of the context. When examining the measures of central tendency for a data set that has a left or negative skew, the clustering of the data set at greater values usually causes the median to be greater than the mean (which is true in this case, as the median is –9 and the mean is –10.16, for the female weight change data).

While histograms show the frequency of the elements of a data set organized into intervals, another way to analyze data sets is through box plots. Box plots organize data sets into quartiles, looking at the spread of data across four equally sized groups of the data set. The data for male and female participants in the weight loss study are shown in figure 6.5.

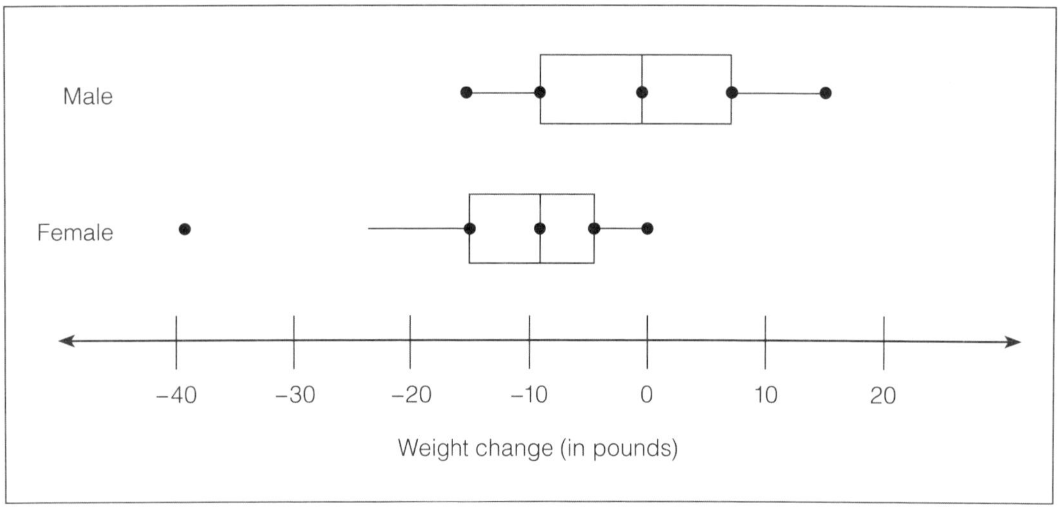

Figure 6.5: Box plots for the weight loss data, organized by gender.

These box plots show that the female data set includes an outlier—a data point that is a distant and extreme value when compared to the rest of the data. The outlier is somewhat hidden in the histogram, represented by the short height of the bar indicating losing between thirty and forty pounds on the far left. In fact, from the histogram, the data set appears contiguous, with the frequency of the data decreasing toward smaller values, or greater weight loss, on the left side of the histogram. From the box plot, you see that there is a large gap between the outlier data value and the majority of the data in the data set. This is one reason why it is important to analyze data sets in multiple ways, as different displays may provide different views of the data. This connects to using multiple representations with functions and learning how each representation provides different benefits in analyzing contextual and mathematical relationships.

How do you address the extreme value in this data set? Outliers often present a challenge when interpreting data. Should you keep the outlier in your analysis or eliminate it? Using histograms to compare data sets, you may choose to eliminate an outlier, as an outlier will elongate the horizontal scale of the histogram, making comparison between two data sets more difficult. From examining the histograms in figures 6.2 (page 137) and 6.4 (page 139), the outlier does not have a substantial impact on the horizontal scale, so you probably would not eliminate it for that reason. With the box plots, the outlier can be clearly represented as an extreme and unique value, so you may choose to retain it in your display and analysis.

Another impact that an outlier can have on data analysis is related to the measure of center. If the outlier is removed from the data set, the overall mean changes from a loss of 5.02 pounds to a loss of 4.85 pounds, and the female participant mean changes from a loss of 10.09 to a loss of 9.80 pounds. Conceptually, the mean can be considered as the balancing point of a data set, creating a measure of center that includes the impact of every data element. However, when there is an outlier, this balancing point is pulled in the direction of the outlier and the mean is less representative of the entire data set, making the mean somewhat misleading as a comparison value for this task. However, the medians for both the whole data set and the female data set do not change if you remove the outlier from the data set, reflecting that an outlier has less effect on the median. When there is an outlier in a data set, you should consider using the median value for comparison, as the value is generally more representative of the entire data set and less sensitive to outliers. The choice of the median as the representative measure of center lessens the impact of an outlier in analyzing the overall data set.

Of course, any analysis in which you choose to exclude data, such as an outlier, needs to clearly explain that choice as well as why that choice was made, particularly in light of the context. With one data element very different than the rest, you can decide to exclude the participant who lost 39 pounds and explain how that one participant was creating a different picture of the entire study, if this description is appropriate.

In this example, you see how to use statistical reasoning using the shape and center of the two sets of data. Predictions were made using the structure of the normal distribution model. Another aspect of this analysis is the need to consider how to properly analyze data based on the context. Originally, the entire set of data was considered, and then the data were explored by gender. This is an important aspect of analyzing real-world data. It may be more appropriate to analyze the program for males differently than for females, as the impact of the program may be different based on gender. Proper analysis of the data requires an understanding of the context and how the data were collected. The context has an impact on analyzing the data and on how the data may apply to the population to which you plan to generalize the results. The question of whether you can use the normal distribution model for each gender impacts your findings, and therefore is an important—and challenging—aspect of your decision making.

The choices you made impacted your analysis of the comparison of the data sets. You were able to consider the values of different displays and the choice of measures of center. In addition, you made a decision about the impact of an outlier and how to address this in your analysis. In mathematizing the situation, your choice of model was based on the data and the context, representing Mathematical Practice 4, "Model with mathematics." These decisions about appropriate choices for displays and their impact on your analysis are similar to the decisions you made in considering different functions to use as parts of a piecewise function in the Facebook users task in chapter 5. This is an important way that mathematical modeling connects across many different mathematical topics.

Understanding Sample Surveys, Experiments, and Observational Studies

Another important component of statistical reasoning at the high school level involves understanding the differences among sample surveys, experiments, and observational studies. Considering which type of data collection model to use and how you will analyze the data collected may be determined by the context. Consider the scenario in figure 6.6.

DO NOW

Imagine you are in charge of research and development at a candy company. Which statistical design (a sample survey, an experiment, or an observational study) would be most appropriate to use in each research project?

- The company asks you to determine whether the candy bar that the company started distributing last year is causing any negative side effects in the people who have eaten it.
- The company asks you to determine if the newly created chewing gum will cause cavities.
- The company wants to know which of its candy is the favorite type across the entire country.

Figure 6.6: Designing sample surveys, experiments, and observational studies task.

How would you design research projects to meet these needs? How would you ensure that your design would produce fair and reliable results? Consider some of the different types of research designs that are available, such as sample surveys, experiments, and observational studies. In a sample survey, you select a subset of your population of interest and ask them questions. If you design the survey well and carefully select the participants, you can use the answers to your sample survey to predict the responses of the entire population. Your goal is to use the responses from a small part of the population to determine how the entire population would respond.

Which of the questions in figure 6.6 do you think could be best answered by using a sample survey? You might design a sample survey to collect information on what type of the company's candy is preferred across the country. You would start by designing questions for a survey that would determine which candy is preferred (and the wording of these questions would be chosen carefully to avoid bias toward one candy over another), then you would create a representative sample of the entire population to respond to these questions. Your sample would need to include a sufficient number of people so that you would have enough participants to account for possible nonresponse bias as not everyone will respond to a survey.

An important factor in selecting your participants is to choose them randomly, and this factor is one way that statistics and probability concepts work together. Random selection allows you to use the rules of probability and statistical analysis to make predictions about a larger population than just the people you ask in your survey. Two key factors in selecting a random sample are that every member of the population has a chance of being included in the sample and that the process of choosing participants is done fairly. How could you have one factor without the other? Consider a case in which every member of a population can be included, but not every member of the population has an *equal* chance of being selected. If you include the entire population of the school but select the participants through their art classes, then this would *not* be an example of random selection. The benefit of random selection is that it tends to produce samples with characteristics similar to the overall population. This is the best way to create a design that will lead to valid inferences.

How do you determine the correct number of participants in your survey? The formula for determining the proper sample size includes concepts that are beyond the scope of this book, such as the degree of confidence you desire in your survey and the maximum margin of error you will allow. The important understanding at this level is that the more accurate you desire your results to be and the better you want your results to reflect the population, the more people you will tend to need to include in your sample. But remember, the more people you survey, typically the greater the cost to conduct your survey.

Now that you have considered how to use a survey, which question would you answer using data from an experiment? You may have chosen an experiment for the second question in figure 6.6. You could create two groups of people with similar characteristics, one group that would be given the new chewing gum and a second group that would not receive the new chewing gum. For an experiment like this, it would be best to randomly select participants, but the challenge is that you are not able to require them to participate. An alternative is to recruit volunteers to participate in the experiment and randomly assign them to either the treatment group (receive the new chewing gum) or the control group (not receive the new chewing gum). Ideally, these two groups have similar characteristics, particularly in their gum-chewing habits. After a certain amount of time, you could compare the number of cavities of the treatment group to the number of cavities of the control group to help determine the potential cavity-causing features of your company's new chewing gum.

Finally, an answer to the question regarding possible negative side effects of the candy bar can use data collected through an observational study. In an observational study, you can examine events that have already happened. You could create two groups of people with similar characteristics, where the only difference is that one group eats the company's candy bars and the other group does not eat the company's candy bars. You would compare the results of these two groups to look for statistical significance. While observational studies are not randomized trials in the sense of an experiment, they can provide insights into the differences between two groups.

These research projects represent the types of designs that high school students learn about and evaluate in statistics and probability. In addition, they learn about features of data sets, data displays, and how concepts related to probability help them make decisions and predictions about the world around them. Being able to assign the proper mathematical model to a given context is an important understanding for students to develop as they make sense of data.

Investigating Probability

Important uses of probability in high school involve simulation and the interpretation of conditional probability. How students interpret simulation and two-way tables, and understand the variance in individual and large-scale predictions, is valuable in making accurate and meaningful predictions from data. This can be seen in predicting outcomes based on given conditions and interpreting two-way frequency tables. For example, consider the challenge provided by the guessing on a multiple-choice task in figure 6.7 (page 144).

DO NOW

There are times when you probably encounter a question on a quiz or test for which you are uncertain of the correct answer. Imagine having to take the quiz below, where you are not provided the answers associated with each question. If each question is worth 5 points, and the entire quiz is worth 20 points, how many points would you expect to get on this quiz?

Today's Quiz

1. $27 - 14 = ?$ a. b. c. d.	2. $56 \div 7 = ?$ a. b. c. d.
3. $42 + 46 = ?$ a. b. c. d.	4. $9 \times 13 = ?$ a. b. c. d.

Figure 6.7: Guessing on a multiple-choice quiz task.

How many points would you expect to get on the quiz? Why? You probably thought that since there are four choices for each problem, you have a one in four chance to get each problem correct, and you would probably earn 5 points out of 20 for the quiz. But how can you simulate whether that is really what you will earn on the quiz? Simulation allows you to model real-world events to predict what may happen in the future, with the caution that you may get different results when you conduct a simulation multiple times. One way to determine your outcome for the multiple-choice quiz would be to create a simulation using random numbers to model quiz results. How would you model a one in four chance of getting the problem correct using random numbers? With a 25 percent chance of getting each question correct, you might choose to use random numbers represented by two digits, with a number between 0 (using 00 to represent a two-digit number) and 24 representing getting a correct answer and a number between 25 and 99 representing an incorrect answer. Choose one of the lines from the random number table (see figure 6.8) and determine your quiz results based on the first four two-digit numbers from the list.

Random Digits

11	16	43	63	18	75	06	13	76	74	26	32	07	51	00	10	43	12	04	18	19	22	89	17	92
21	21	59	17	91	76	83	15	86	78	87	05	43	16	87	93	20	54	36	85	19	73	20	84	68
10	43	84	44	82	66	55	83	76	49	08	88	29	08	70	12	46	24	18	10	01	80	60	29	77
36	79	22	62	36	33	26	66	65	83	60	88	19	73	95	20	46	13	67	42	02	85	25	05	64
73	94	40	47	73	12	03	25	14	14	82	38	43	83	70	00	24	98	07	09	72	60	56	74	97

Source: National Institutes of Standards and Technology, n.d.

Figure 6.8: Random number table.

If you used the first line from the table, your first four two-digit random numbers are 11–16–43–63, and based on the described assignment, you would get two answers correct since there are two numbers between 00 and 24, for 10 points out of 20 points on the quiz. The next set of four values is 18–75–06–13, which provides a simulation where you get three of the four answers correct. Other sets give other results, and if you do enough simulations, you will see results similar to your expectation of getting the correct answer about 25 percent of the time, or earning five points out of twenty most of the time. This is important in understanding statistical reasoning, variability, and simulation. Results will be different with different simulations, and predicting one specific event is often very difficult to do. The understanding of variability and the interpretation of results is also important when students analyze two-way frequency tables, which is evident in one of the classroom videos presented later in this chapter. These understandings lead to creating mathematical representations that demonstrate how randomness can be used to model real-world events, an example of Mathematical Practice 4, "Model with mathematics."

Exploring Scatterplots

Students explore and measure bivariate relationships in many ways. You have explored examining relationships involving different types of functions, and this understanding can be extended to explore the relationship between variables in a scatterplot. You can think of the visual provided with a scatterplot in terms of how the points are situated on the graph—the scatter of the plot provides a visual of the strength of the relationship between the variables. What can you learn from a scatterplot using the data in figure 6.9 (page 146)?

How did you make your prediction? How did you use your scatterplot to help? In creating your scatterplot, you may have noticed that there is a pattern in the data (see figure 6.10, page 147).

A key feature of high school mathematics is the ability to use a wide range of functions to assist in analyzing real-world situations. In the middle grades and early high school mathematics, the focus is on linear relationships. As students develop deeper understandings of function types, they learn to use additional functions, including exponential and quadratic functions. You experienced these functions in exploring the Facebook users task in chapter 5. How will you determine which function is the best fit for the patterns in the insurance cost data? You probably noticed that each of the four sets of data is increasing, but none of them are increasing at a constant rate of change, so they are not represented by linear patterns. What type of function is the best fit for each set of data? The correlation coefficient provides you a way to determine which function is the best fit. The correlation coefficient measures the degree of dependence of one variable on another or how well the two variables relate to one another, relative to the function type selected. The closer the correlation coefficient is to the value of 1 (or –1), the better the fit of the function to the data. You can use technology to determine each of the potential relationships—linear, exponential, and quadratic—and the measure of the correlation coefficient, which is represented by R (see figure 6.11, page 148).

Which function is the best fit for these data? Because the correlation coefficients are closest to 1 for the exponential relationship for males (R = 0.992) and the quadratic relationship for females (R = 0.988), you know that they offer the best fit for each data set. Using these functions, the prediction for the cost of life insurance for a 75-year-old male is about $113.01 a month and for a 75-year-old female is about $71.03 a month. Which relationship is the best for the long-term-care insurance rate (see figure 6.12, page 149)?

DO NOW

Heather collected data on life insurance and long-term-care insurance by gender from Forever Life Insurance. She collected the following information.

Life Insurance Rates—$1,000,000 Coverage, Monthly Payment by Age

Age	Male	Female
25	$9.45	$8.81
30	$9.64	$8.90
35	$9.72	$8.92
40	$10.88	$10.06
45	$13.77	$12.30
50	$18.42	$15.48
55	$26.15	$20.13
60	$39.30	$27.78
65	$61.10	$40.56
70	$94.57	$59.98

Long-Term-Care Insurance—$100,000 Total Benefit, Monthly Payment by Age

Age	Male	Female
25	$45.98	$78.45
30	$45.98	$78.45
35	$45.98	$78.45
40	$51.44	$87.64
45	$61.98	$100.43
50	$70.15	$114.12
55	$80.78	$132.98
60	$102.43	$162.97
65	$127.67	$198.45
70	$168.32	$256.34

Create two scatterplots—one of the life insurance data and one of the long-term-care insurance data—and describe the relationship for the monthly payments for males and females. What would you predict the cost to be for the monthly payment for a 75-year-old male and female for life insurance and long-term-care insurance? What might be responsible for the differences in the costs for males and females?

Figure 6.9: Scatterplot task.

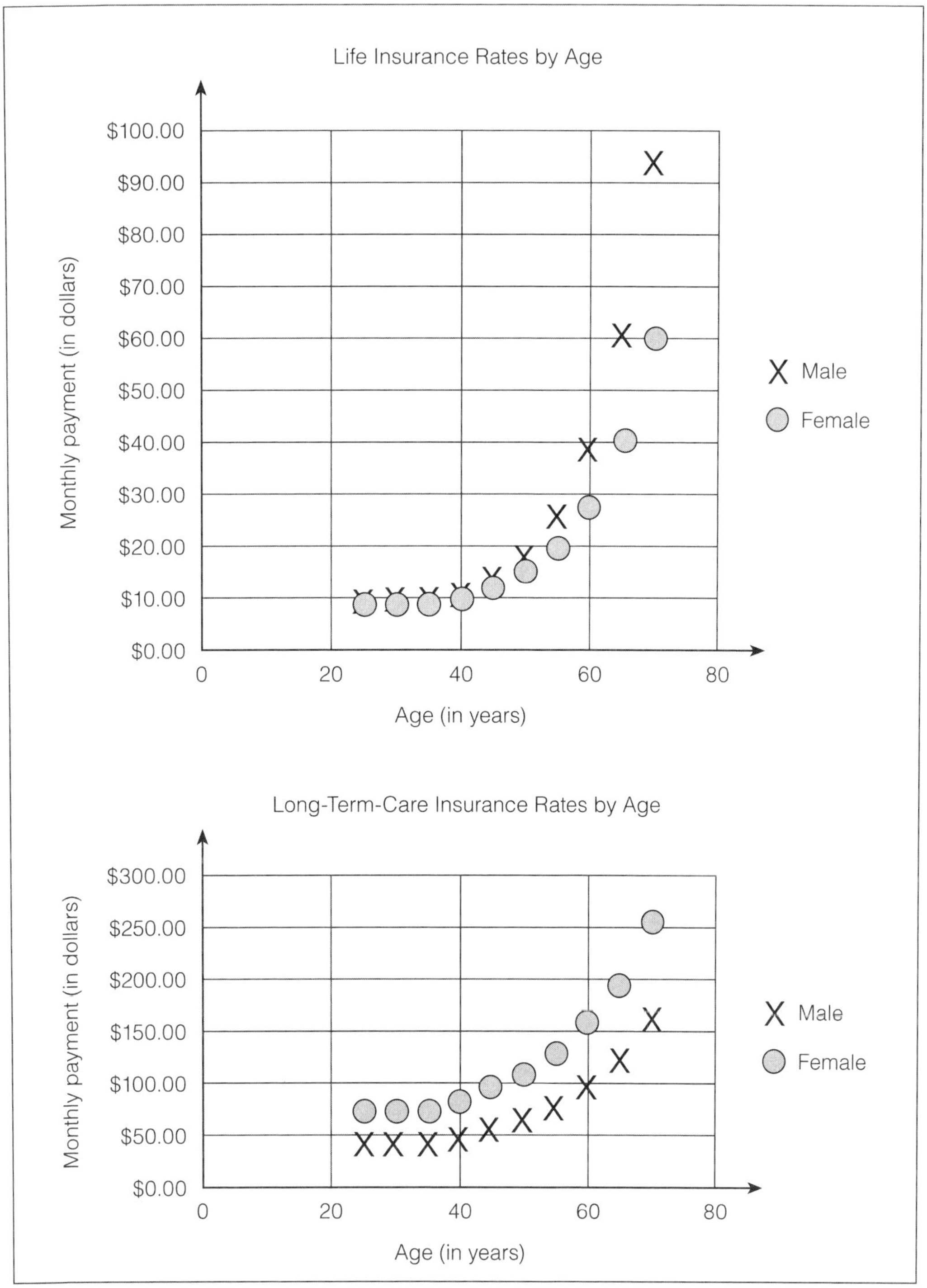

Figure 6.10: Scatterplots of the insurance data.

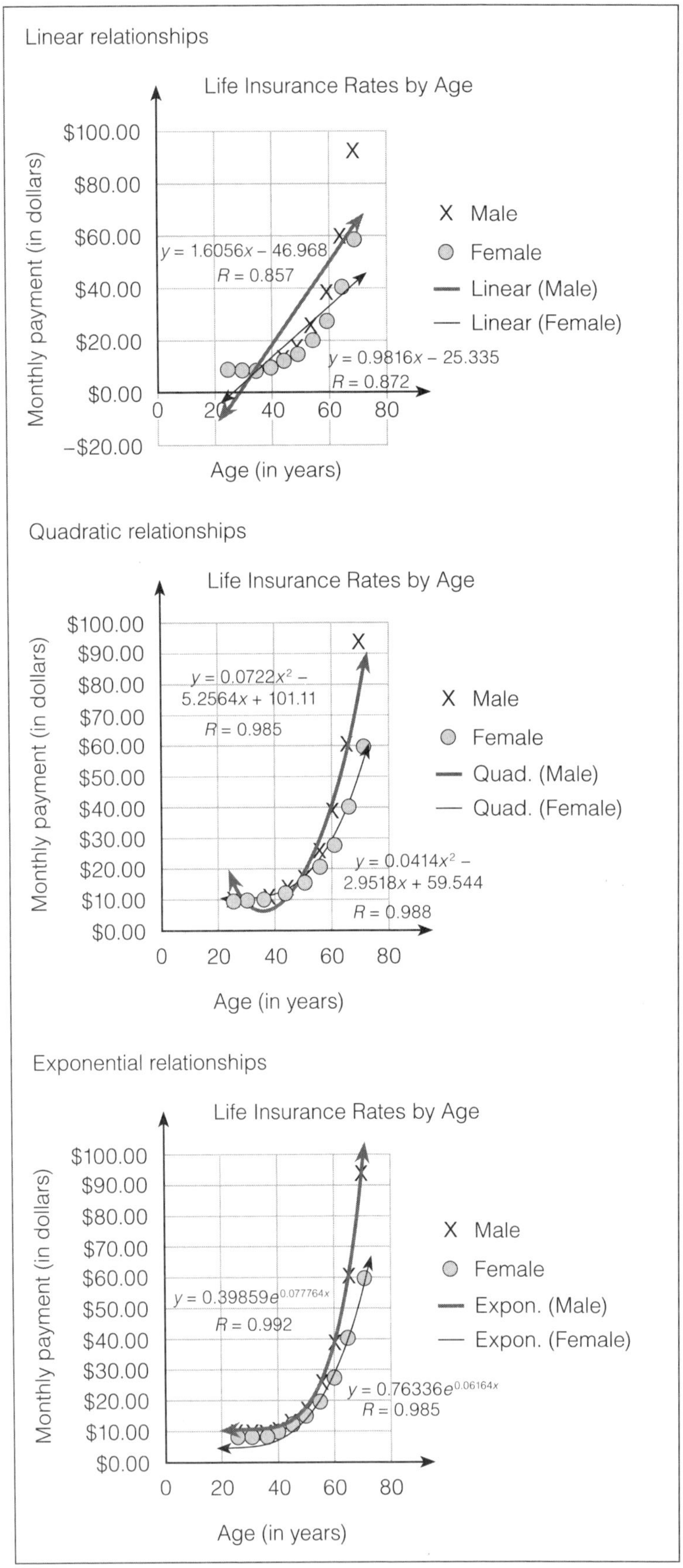

Figure 6.11: Functional relationships for the life insurance cost data.

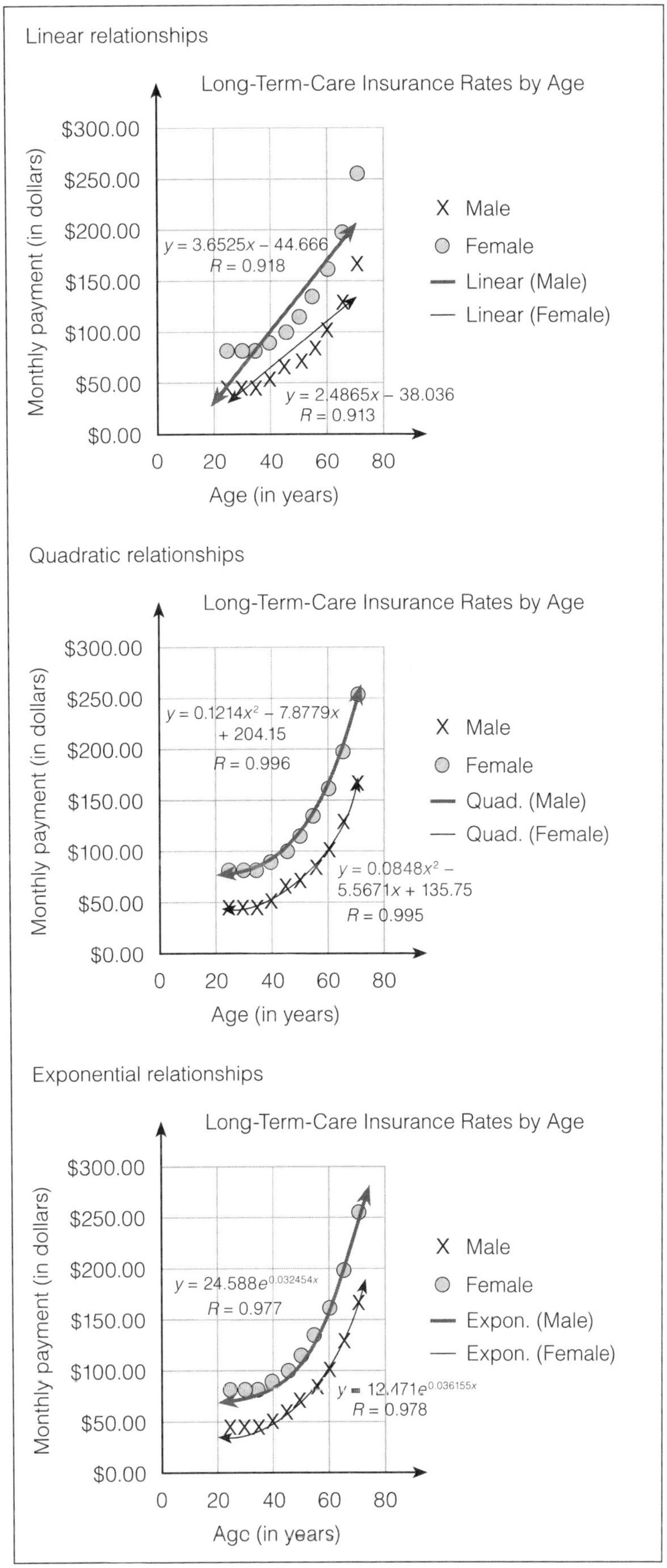

Figure 6.12: Functional relationships for the long-term-care insurance cost data.

This time the quadratic relationship shows the highest correlation coefficient for both males and females. The functions of best fit are $y = 0.0848x^2 - 5.5671x + 135.75$ for males and $y = 0.1214x^2 - 7.8779x + 204.15$ for females. This provides a monthly payment of about \$195.22 for a 75-year-old male and \$269.18 for a 75-year-old female. An important feature of all of these analyses is that when you model the relationship using the quadratic relationship, you need to be cautious of the domain with which you work. Note that the quadratic relationships have positive y-intercepts that are greater than some of the values in the given data set. In the context of these problems, it would mean that insurance rates would be higher for young children, as all four shows a vertex between the ages of 30 and 40. This does not seem reasonable for the provided context. Since the cost of insurance would increase on both sides of this vertex in all four quadratic models, these models may not be effective predictors for the costs for younger people. Whether this is true is not evident in the given data, so it is important that you are careful in your predictions. Think about how the domain may need to be limited in order to make predictions that are meaningful to the context of this task.

Will your prediction be accurate? There are two factors to consider here. One factor is variability. While the data do follow a pattern, there will be some variability in the data. This means that predictions based on your best-fit function may not always be exactly correct. Your understanding of variability allows you to realize that while your prediction will generally be correct, there will always be a degree of variance when you make an individual prediction, so there may be some difference between your prediction and the actual value.

Another factor involves the distinction between correlation and causation. Why is this value called the *correlation* coefficient rather than the *causation* coefficient? It may appear that this value can be used both for correlation and causation. While the correlation coefficient does show how well the data fit to a given function, it does not indicate whether one variable actually causes the other. Causation is determined by the design of the experiment, not the relationship of the data. Think back to earlier in the chapter about the relationship between buying an ice cream cone and getting a sunburn. You need to remember that your prediction is based on a correlation, whether or not a causal effect is present.

Did you also notice that the monthly cost of life insurance is greater for males but long-term-care insurance is greater for females? What does that mean in the context of the task? Insurance companies use projected life expectancy to help determine the cost of insurance, which is based on health, age, and family history, among other things. What do you deduce about the life expectancy for males and females based on this information? The U.S. Census Bureau (2011) estimates that a child born in 2015 has a life expectancy of 78.9 years—76.4 for males and 81.4 for females. How is this information reflected in the collected life insurance and long-term-care insurance rates by gender from the Forever Life Insurance Company? Since females in the United States tend to live longer than males, you would expect that they would pay less for life insurance (because they tend to pay for a longer period of time) and they would pay more for long-term-care insurance (because they tend to need it for a longer period of time). This is an important feature of understanding mathematics—knowing how the mathematics models relevant characteristics of the real-world context. The ability to make sense of the features of the context and the mathematical understandings of the graphs is an important demonstration of Mathematical Practice 2, "Reason abstractly and quantitatively."

The Classroom

The videos for this chapter represent classroom examples of the same type of statistical reasoning that you have experienced in your tasks. You will explore how students make sense of this reasoning in a middle grade and a high school classroom. The first video engages students with a data set and its accompanying dot plot. Students begin the lesson by estimating the number of jelly beans in a jar as a way to consider measures of center and to connect to a data set they will analyze. You are encouraged to watch the video in its entirety before proceeding.

www.solution-tree.com/Determining_the_Appropriate_Measure_of_Center

How do the students work to make sense of the different measures of center? How does the teacher respond to student ideas?

What did you notice about the opening of the task? Asking students to determine their own estimate for the number of jelly beans in the container allows them to connect to the context by making their own conjectures. In some lessons, students struggle to make sense of the activity because they fail to grasp the context in which the task is set. By providing this introduction to the task, all students experience how the data may have been collected and what the data set represents. This provides a common background for students so that everyone starts with an understanding of the context, which is achieved through the level of the layers of facilitation, "I facilitate the whole class."

How do students understand the role of statistics in making sense of the task? Some students struggle to blend their personal experiences and views with analysis of the given data set. When asked for a number that would represent the entire data set, one student's response is 109, the smallest number in the given data set. His justification comes not from statistical reasoning using the data set, but rather in regards to the size of the container and his group's estimates. How would you address this response? The teacher does not identify the response as being incorrect, but rather identifies how the student determined his response based on his group's estimates.

The class next discusses the median and the mean as possible choices for a single value to represent the data set. Notice how the teacher asks questions of the students to have them provide the definitions of both median and mean and shares with them the value for each measure. Notice how this ties to the learning goal of analyzing the data set. The key here is not to calculate the median or the mean, but rather to have the students provide the definition of these terms and discuss the role of mean and median in representing this specific data set. With that learning goal in mind, the teacher chooses not to take time in this task to calculate the median or the mean, but rather to discuss how to determine these values and then provide them. This keeps the focus on the concepts rather than the procedures.

How are the measures of central tendency addressed in the lesson? After the class discusses the meaning of the mean and the median, a student discusses how the differences in particular numbers in the data set can have an impact on these values. The teacher responds by asking the student for a specific number that appears to be different than the rest. While the term *outlier* is not a requirement for student understanding at this grade level, providing students information within the context of the lesson supports their understanding in later coursework. Near the end of the lesson, the teacher has a student connect his original estimate to the given data set and the student links it to the concept of outlier.

In the second video, high school students are examining a two-way table and considering aspects of conditional probability. Please watch this video before proceeding.

www.solution-tree.com/Interpreting_Two-Way_Frequency_Tables

In this video, students are investigating a two-way table to determine if two different conditional probabilities are the same no matter the order of the conditions—the probability of having a job if you own a car and the probability of owning a car if you have a job. The two-way table is set up in a way that these two conditional probabilities are equal. What do you notice about how the students approach the task? How does the teacher encourage students to engage in the task?

The challenge for students is to see beyond the specific numbers and determine if the probabilities will be equal if the numbers are different. Notice how the groups discuss their thinking about the task. One group considers only the numbers in the table and makes their conclusions based on those numbers. Another group considers how if the numbers change, the relationships might be different and the two probabilities would not be the same. Notice how the questioning of the teacher moves from student to student in each group to determine if the students agree with one another. The teacher is seeing the degree to which the students are sharing their thinking and making sense of the problem. These students are engaging in Mathematical Practice 1, "Make sense of problems and persevere in solving them."

TQE Process

Now it might be helpful to watch the first video (page 151) for a second time, paying close attention to the tasks, questioning, and opportunities to collect evidence of student learning.

The TQE process can help you frame your observations. Teachers who have a deep understanding of the mathematics they teach:

- Select appropriate *tasks* to support identified learning goals
- Facilitate productive *questioning* during instruction to engage students in Mathematical Practices
- Collect and use student *evidence* in the formative assessment process during instruction

The *task* chosen for this lesson helps students connect statistical concepts to a tangible event. The numbers included in the task were carefully selected in order to build the understanding of the concepts selected. For example, the data set specifically included an outlier value so student discussion would need to consider how to address this value. The impact of the outlier on the values of the mean and the median, and how to address the outlier in the analysis, can only be considered if the value is a part of the data set. It is important in the selection or creation of tasks that the data set supports the depth of understanding desired for the learning goal. You might notice that the teacher is intentional about allowing the students to make sense of the task themselves. Initially, students discuss their own estimates. This allows them to make sense of the context in very personal and potentially different ways. Once these estimates are made, the teacher facilitates the students to think about the learning goal by asking about having one number represent the group of estimates. The task directly supports this learning goal and was chosen to provide opportunities to compare the measures of central tendency and to introduce the link between these measures and the concept of outliers. This promotes productive talk among the students by helping them connect to the data on the dot plot. The teacher helps students stay engaged in the group conversation and helps students make sense of each other's work. The teacher uses questions to guide students to consider which measure of center would be better for the given data set without telling the students the answers, thus using the task to support the learning goal.

Notice how the teacher uses *questioning* and the conversations in the class to assess how students are making sense of the task and of the mathematical content. He asks questions in order to determine whether students understand the definition of both mean and median and how to apply those meanings in the context of the task. He allows students to extend their thinking when a student states that his estimate of 40 would be an outlier for the given data set. When this same student offers the data element of 109 to represent the data set, the teacher guides the discussion to include other students and other ideas. In order to check that this student learned about measures of central tendency during the lesson, the teacher returns to that same student at the end of the lesson to check for understanding. Through strategic use of questions and the selection of students to answer those questions, the teacher uses a formative assessment process to support the development of understanding.

The students provide *evidence* of their learning as they engage in the lesson. Students connect their own estimates to the estimates in the given data set. They begin by making sense of how the data are collected, individually and as a class, and then move to analyze a display of a different set of data. After they make sense of the given data, they link back to their estimates and integrate the two data sets together, demonstrating how one of their estimates would act as an outlier in the given data set. By reasoning with actual estimates as well as a given set of data, and connecting to the context in multiple ways, these students engage in Mathematical Practice 2, "Reason abstractly and quantitatively."

You are encouraged to take some time to reflect on your own teaching. In what ways can you allow students the time and opportunity to make sense of the concepts you are teaching? How can you engage your students in meaningful conversations with one another and establish the expectation that they work together to develop shared understanding of the concepts?

The Response

One of the major challenges with statistical reasoning for high school students is the role of randomness in the world of statistics and probability. When percentages are included in the likelihood of the outcome of events, the role of individual outcomes versus understanding the overall group behavior can prove challenging. For example, when students hear that someone has flipped a coin ten times and gotten heads each time, what do they think? Do they believe that getting a tails on the next coin flip is equally likely to getting heads? Or do they think that tails will be more likely, abiding by some type of law of averages, to make up for the long run of consecutive heads? This is one of the key challenges in understanding probability—the fact that the tools used to explore probability do not have a memory. The outcomes of flipping a coin are not impacted by what has happened before. Making accurate predictions relies on using the given data and not allowing predictions and conclusions to be influenced by experiences or impressions.

Another area where students may struggle is assessing the impact of bias in survey samples, experiments, and observation studies. In order to determine a fair sample, one has to ensure that all possible participants have an equal chance of participating. However, getting that fair sample can sometimes be difficult to define and difficult to accomplish if participants have the opportunity to opt out of participating. In addition, some samples that appear to be unbiased can actually contain hidden biases for which it is difficult to control. How students think about countering bias, and how to anticipate the effects of these biases, is an important feature of statistical reasoning.

Reflections

1. What do you feel are the key points of this chapter?
2. What challenges might you face when implementing the key ideas from this chapter? How will you overcome them?
3. What are the important features for developing an understanding of statistics and probability, and how will you ensure your instruction embeds the support needed for these features?
4. Select a recent lesson you have taught or observed focused on statistical reasoning. Relate this lesson to the TQE process.
5. What changes will you make to your planning and instruction based on what you read and considered from this chapter?

EPILOGUE

Next Steps

An important role of mathematics teachers is to help students understand mathematics as a focused, coherent, and rigorous area of study, regardless of the specific content standards used. To teach mathematics with such depth, you must have a strong understanding of mathematics yourself as well as a myriad of teaching strategies and tools with which to engage students. Hopefully, by providing the necessary knowledge, tools, and opportunities for you to become a *learner* of mathematics once more, this book has empowered you to fill this role.

Now what? How do you take what you learned from *doing* mathematics and make good use of it as the *teacher* of mathematics?

Our position is that you first need to apply what you learned to your lesson planning. Are you planning for instruction that focuses on teaching concepts before procedures? How is your planning aligned to developing learning progressions? How will you ensure that your lessons do not end up as a collection of activities? What follows are strategies that will help you use what you experienced as learners and apply it to what you do as teachers.

Focus on Content

At the heart of meaningful mathematics experiences is mathematics content. A focus on content addresses the *what* of mathematics instruction. What is the mathematical idea or concept you want students to develop and learn as a result of the lesson you facilitate? With your collaborative team, discuss the content that will best serve your students as you progress through the school year. Everything that happens in a mathematics lesson—every task, every activity, every question, every element of the formative assessment process—provides an opportunity to strengthen students' understanding of mathematics content. Thus, it is important for you and your team to engage in collaborative planning about the mathematics content of a unit before the unit begins. Having made sense of mathematics for teaching provides you with a focus on content that will help you and your students have more meaningful and productive experiences with mathematics.

Select Good Tasks

As you've seen throughout this book, your focus on content is revealed in the tasks you select for students to engage in during instruction, so be sure to address this element of instruction during planning. Good tasks are those that support students in learning meaningful mathematics (concepts and, when appropriate, procedures). Other byproducts of good tasks include students engaging in meaningful discourse, developing critical thinking, having multiple ways of representing their thinking (definitions, equations, drawings, and so on), and building fluency (choosing strategies that are most efficient for a given task) in mathematics.

Good tasks also support students in developing proficiency with the Mathematical Practices, thereby supporting students' development in mathematics that will last beyond their current grade of study. As you teach, you have the opportunity to engage in the formative assessment process through the use of good tasks. By selecting good tasks, you set the stage for your students to develop strong conceptual understandings of mathematics. We modeled this for you through the tasks provided in this book.

Align Instruction With the Progression of Mathematics

Knowing how mathematics progresses within and across grades is a valuable asset for planning mathematics lessons. As students develop mathematically, you want their classroom experiences to be aligned with how the mathematics should progress. These understandings are important in both planning and implementing instructional processes. In this book, we've facilitated the alignment of instruction with the progression of mathematics within and across grades.

Build Your Mathematics Content Knowledge

Throughout this book, we have addressed mathematics content that is meaningful for your grade band. We have also addressed the content from the perspective that teachers need to facilitate meaningful mathematics instruction. In each chapter, we encouraged you to engage with the mathematics for the purpose of building your mathematics content knowledge by doing mathematics. Our aim was to provide you the support you needed for your understanding of mathematics so that you could subsequently engage your students in similar ways.

Observe Other Teachers of Mathematics

Making Sense of Mathematics for Teaching High School has provided you the opportunity to observe mathematics teaching in action. Through videos and examples, you were able to see students engaged in meaningful mathematics based on rich tasks. We hope this provided an opportunity for you to nurture discourse within your collaborative teams about teaching and learning mathematics. For each video, our aim was to guide you to examine the teacher moves and student exchanges that support the students' learning of mathematics. We are confident that each video provides nuggets of insight and confirmation that will help you clarify your thinking about and improve your teaching of mathematics. We encourage you to continue this practice by observing your fellow teachers when possible and inviting them to observe you.

Respond Appropriately to Students' Struggles With Mathematics

Students' progress in mathematics is often met with some level of struggle with understanding mathematics concepts and applying procedures. Some students even struggle with the language (words, symbols, and so on) of mathematics. Some misunderstandings are unintentionally perpetuated when teachers do not deeply understand mathematics (such as, subtracting always makes smaller, or multiplication always makes larger). By preparing for common misunderstandings and errors, you will be better able to help students successfully engage with mathematics and overcome barriers to understanding. Using the formative assessment process to determine what students know and do not know provides opportunities for you and your collaborative team to reflect together to improve instruction.

Now What?

How has *Making Sense of Mathematics for Teaching High School* helped prepare you for your next steps? How will you use the TQE process to inform your practice?

- How will you select appropriate *tasks* to support identified learning goals?
- How will you facilitate productive *questioning* during instruction to engage students in the Mathematical Practices?
- How will you collect and use student *evidence* in the formative assessment process during instruction?

In responding to these questions within your collaborative team, your focus should also include the four critical, guiding questions of PLCs (DuFour et al., 2010).

1. What do we want students to learn and be able to do?
2. How will we know if they know it?
3. How will we respond if they don't know it?
4. How will we respond if they do know it?

Now that you have reached the conclusion of this book, we also ask you to respond to these three questions in the spirit of continuing the reflective process.

1. What do you know now that you did not know before interacting with this book?
2. What do you still need to learn now that you have completed this book?
3. How will you obtain the knowledge you still need?

Appendix: Hypothetical Weight Loss Study Data

Gender	Pre-Treatment Weight	Post-Treatment Weight	Difference
M	248	258	10
F	145	135	–10
M	272	268	–4
F	133	123	–10
M	186	197	11
M	271	281	10
M	206	201	–5
M	240	248	8
F	174	163	–11
F	153	150	–3
F	179	177	–2
M	259	250	–9
M	189	204	15
F	161	157	–4
M	249	238	–11
M	204	197	–7
M	188	198	10
M	263	272	9

continued →

Gender	Pre-Treatment Weight	Post-Treatment Weight	Difference
F	182	168	−14
F	136	135	−1
F	167	158	−9
M	270	267	−3
F	146	141	−5
M	213	198	−15
F	132	123	−9
F	174	174	0
F	190	168	−22
F	184	172	−12
M	176	162	−14
M	178	174	−4
M	219	205	−14
F	179	166	−13
M	243	240	−3
F	173	153	−20
F	164	164	0
M	207	220	13
F	151	146	−5
M	210	210	0
M	274	263	−11
F	180	158	−22
F	142	123	−19
M	231	226	−5
M	199	214	15
F	151	131	−20
F	152	131	−21
F	181	161	−20
F	141	132	−9

Gender	Pre-Treatment Weight	Post-Treatment Weight	Difference
M	252	249	–3
M	239	247	8
F	156	149	–7
M	196	186	–10
F	163	162	–1
M	205	195	–10
F	180	178	–2
F	147	147	0
F	158	149	–9
M	185	171	–14
F	165	162	–3
F	138	133	–5
F	134	113	–21
F	166	162	–4
M	199	206	7
F	170	167	–3
M	190	188	–2
M	205	210	5
F	183	178	–5
F	161	154	–7
F	132	121	–11
M	210	207	–3
M	209	221	12
F	141	102	–39
M	259	244	–15
F	162	148	–14
F	160	152	–8
M	232	228	–4
M	237	238	1

continued →

Gender	Pre-Treatment Weight	Post-Treatment Weight	Difference
M	262	251	−11
F	181	180	−1
M	196	184	−12
M	188	195	7
F	189	170	−19
M	251	248	−3
F	177	168	−9
M	198	203	5
F	145	144	−1
F	195	179	−16
M	177	180	3
F	180	177	−3
M	253	257	4
F	133	119	−14
F	145	139	−6
M	205	211	6
M	266	255	−11
F	180	175	−5
F	169	164	−5
F	147	127	−20
M	214	200	−14
M	187	185	−2
M	223	225	2
M	206	199	−7
M	198	192	−6
M	178	177	−1
M	267	256	−11
M	257	253	−4
F	144	130	−14

Gender	Pre-Treatment Weight	Post-Treatment Weight	Difference
F	194	193	–1
F	143	129	–14
M	239	247	8
M	232	230	–2
M	197	194	–3
F	146	123	–23
M	189	188	–1
F	192	183	–9
M	274	284	10
M	232	219	–13
M	227	233	6
M	238	249	11
F	184	176	–8
F	174	173	–1
F	178	167	–11
F	149	129	–20
M	235	248	13
F	183	171	–12
F	143	137	–6
M	230	226	–4
F	164	156	–8
F	150	144	–6
F	185	176	–9
M	196	208	12
M	201	194	–7
M	240	247	7
F	145	135	–10
M	262	271	9
M	185	191	6

continued →

Gender	Pre-Treatment Weight	Post-Treatment Weight	Difference
F	159	138	−21
F	160	142	−18
M	244	258	14
F	186	170	−16
M	271	256	−15
F	132	125	−7
M	244	248	4
F	191	177	−14
M	188	202	14
M	221	224	3
F	158	141	−17
M	238	237	−1
M	256	264	8
M	243	252	9
M	219	229	10
F	177	171	−6
F	132	131	−1
F	172	168	−4
M	247	257	10
F	155	151	−4
M	269	282	13
M	243	228	−15
F	165	145	−20
F	151	138	−13
M	228	222	−6
F	188	174	−14
F	193	183	−10
F	153	142	−11
M	245	252	7

Gender	Pre-Treatment Weight	Post-Treatment Weight	Difference
F	153	130	−23
F	170	165	−5
M	193	178	−15
M	208	213	5
M	259	264	5
M	220	207	−13
F	153	152	−1
M	190	198	8
M	261	272	11
F	180	179	−1
M	183	169	−14
M	250	248	−2
F	133	110	−23
M	250	241	−9
F	138	134	−4
F	174	158	−16
F	171	162	−9
M	201	205	4
M	224	209	−15
F	155	150	−5
F	132	131	−1
F	154	146	−8
F	171	169	−2
M	212	202	−10
M	180	195	15
F	135	120	−15
M	257	268	11
M	189	204	15
F	181	162	−19

continued →

Gender	Pre-Treatment Weight	Post-Treatment Weight	Difference
M	222	215	–7
F	165	155	–10
F	152	149	–3
M	258	264	6
F	151	128	–23
M	202	192	–10
F	175	161	–14
F	178	178	0

References

Associated Press. (2012, October 23). *Number of active users at Facebook over the years.* Accessed at http://finance.yahoo.com/news/number-active-users-facebook-over-years-214600186--finance.html on November 14, 2015.

Ball, D. L., Thames, M. H., & Phelps, G. (2008). Content knowledge for teaching: What makes it special? *Journal of Teacher Education, 59*(5), 389–407.

Dixon, J. K., Adams, T. L., & Nolan, E. C. (2015). *Beyond the Common Core: A handbook for mathematics in a PLC at Work: Grades K–5.* T. D. Kanold (Ed.). Bloomington, IN: Solution Tree Press.

DuFour, R., DuFour, R., Eaker, R., & Many, T. (2010). *Learning by doing: A handbook for Professional Learning Communities at Work* (2nd ed.). Bloomington, IN: Solution Tree Press.

Facebook. (2013). *Facebook annual report 2013.* Accessed at http://files.shareholder.com/downloads/AMDA-NJ5DZ/1008323968x0x741493/EDBA9462-3E5E-4711-B0B4-1DFE9B541222/FB_AR_33501_FINAL.pdf on November 14, 2015.

Fisher, D., & Frey, N. (2003). Writing instruction for struggling adolescent readers: A gradual release model. *Journal of Adolescent and Adult Literacy, 46*(5), 396–405.

Garcia, G. G., & Cox, R. (2010). "Graph-as-picture" misconceptions in young students. In A. K. Goel, M. Jamnik, & N. H. Narayanan (Eds.), *Lecture notes in artificial intelligence* (Vol. 6170, pp. 310–312). Heidelberg, Germany: Springer.

Janvier, C. (1981). Use of situations in mathematics education. *Educational Studies in Mathematics, 12*(1), 113–122.

Kilpatrick, J., Swafford, J., & Findell, B. (Eds.). (2001). *Adding it up: Helping children learn mathematics.* Washington, DC: National Academies Press.

Kisa, M. T., & Stein, M. K. (2015). Learning to see teaching in new ways: A foundation for maintaining cognitive demand. *American Educational Research Journal, 52*(1), 105–136.

Leinhardt, G., Zaslavsky, O., & Stein, M. K. (1990). Functions, graphs, and graphing: Tasks, learning, and teaching. *Review of Educational Research, 60*(1), 1–64.

National Council of Supervisors of Mathematics. (2014). *It's TIME: Themes and imperatives for mathematics education.* Bloomington, IN: Solution Tree Press.

National Council of Teachers of Mathematics. (2014). *Principles to actions: Ensuring mathematical success for all.* Reston, VA: Author.

National Governors Association Center for Best Practices & Council of Chief State School Officers. (2010). *Common Core State Standards for mathematics.* Washington, DC: Authors. Accessed at www.corestandards.org/assets/CCSSI_Math%20Standards.pdf on January 31, 2015.

National Institutes of Standards and Technology. (n.d.). *Appendix B: Random number tables.* Accessed at www.nist.gov/pml/wmd/pubs/upload/AppenB-HB133-05-Z.pdf on December 3, 2015.

Nolan, E. C., Dixon, J. K., Roy, G. J., & Andreasen, J. B. (2016). *Making sense of mathematics for teaching grades 6–8.* Bloomington, IN: Solution Tree Press.

Statista. (2015). *Number of monthly active Facebook users worldwide.* Accessed at www.statista.com/statistics/264810/number-of-monthly-active-facebook-users-worldwide on November 14, 2015.

Stein, M. K., & Smith, M. S. (1998). Mathematical tasks as a framework for reflection: From research to practice. *Mathematics Teaching in the Middle School, 3*(4), 268–275.

U.S. Census Bureau. (2011). Section 2: Births, deaths, marriages, and divorces. In *Statistical abstract of the United States: 2012* (pp. 63–98). Washington, DC: Author. Accessed at www.census.gov/prod/2011pubs/12statab/vitstat.pdf on November 30, 2015.

Index

G

H

I

L

M

N

O

P

Q

R

S

T

V